Glaciers of Alaska

By Bruce Molnia, Ph.D.

ALASKA GEOGRAPHIC® / Volume 28, Number 2 / 2001

To teach many more to better know and more wisely use our natural resources...

EDITOR
Penny Rennick

PRODUCTION DIRECTOR
Kathy Doogan

ASSOCIATE EDITOR
Susan Beeman

MARKETING DIRECTOR
Mark Weber

ADMINISTRATIVE ASSISTANT
Melanie Britton

ISBN: 1-56661-055-9

PRICE TO NON-MEMBERS THIS ISSUE: $23.95

PRINTED IN U.S.A.

POSTMASTER:
Send address changes to:

ALASKA GEOGRAPHIC®
P.O. Box 93370
Anchorage, Alaska 99509-3370

COVER: *Scott Christy explores Excelsior Glacier's* terminus *on Big Johnstone Lake. Excelsior descends from the Sargent Icefield 25 miles east of Seward. (Curvin Metzler)*

PREVIOUS PAGE: *Common on Alaska's peaks,* cirque glaciers *cling to granite faces of the Cathedral Spires in Denali National Park. (Fred Hirschmann)*

FACING PAGE: *Numerous medial* moraines *follow Barnard Glacier's flow. One of many* valley glaciers *that cut through the St. Elias Mountains, its name derives from Edward Chester Barnard, USGS topographer in charge of boundary surveys for the United States and Canada from 1903 to 1915. (AGS file photo)*

BOARD OF DIRECTORS
Kathy Doogan
Carol Gilbertson
Penny Rennick

Robert A. Henning, **PRESIDENT EMERITUS**

ALASKA GEOGRAPHIC® (ISSN 0361-1353) is published quarterly by The Alaska Geographic Society, 639 West International Airport Rd. #38, Anchorage, AK 99518. Periodicals postage paid at Anchorage, Alaska, and additional mailing offices. Copyright © 2001 The Alaska Geographic Society. All rights reserved. Registered trademark: Alaska Geographic, ISSN 0361-1353; key title Alaska Geographic. This issue published June 2001.

THE ALASKA GEOGRAPHIC SOCIETY is a non-profit, educational organization dedicated to improving geographic understanding of Alaska and the North, putting geography back in the classroom, and exploring new methods of teaching and learning.

MEMBERS RECEIVE *ALASKA GEOGRAPHIC*®, a high-quality, colorful quarterly that devotes each issue to monographic, in-depth coverage of a specific northern region or resource-oriented subject. Back issues are also available (see page 112). Membership is $49 ($59 to non-U.S. addresses) per year. To order or to request a free catalog of back issues, contact: Alaska Geographic Society, P.O. Box 93370, Anchorage, AK 99509-3370; phone (907) 562-0164 or toll free (888) 255-6697, fax (907) 562-0479, e-mail: akgeo@akgeo.com. A complete listing of our back issues, maps and other products can also be found on our website at www.akgeo.com.

SUBMITTING PHOTOGRAPHS: Those interested in submitting photos for possible publication should write or refer to our website for a list of upcoming topics or other photo needs and a copy of our editorial guidelines. We cannot be responsible for unsolicited submissions. Please note that submissions must be accompanied by sufficient postage for return by priority mail plus delivery confirmation.

CHANGE OF ADDRESS: When you move, the post office may not automatically forward your *ALASKA GEOGRAPHIC*®. To ensure continuous service, please notify us at least six weeks before moving. Send your new address and membership number or a mailing label from a recent issue of *ALASKA GEOGRAPHIC*® to: Address Change, Alaska Geographic Society, Box 93370, Anchorage, AK 99509-3370.

If your issue is returned to us by the post office because it is undeliverable, we will contact you to ask if you wish to receive a replacement for a small fee to cover the cost of additional postage to reship the issue.

The Library of Congress has cataloged this serial publication as follows:

Alaska Geographic. v.1-
  [Anchorage, Alaska Geographic Society] 1972-
  v. ill. (part col.). 23 x 31 cm.
  Quarterly
  Official publication of The Alaska Geographic Society.
  Key title: Alaska geographic, ISSN 0361-1353.

  1. Alaska—Description and travel—1959-
    —Periodicals. I. Alaska Geographic Society.

F901.A266 917.98'04'505 72-92087
Library of Congress 75[79112] MARC-S.

COLOR SEPARATIONS: Graphic Chromatics
PRINTING: Banta Publications Group / Hart Press

ABOUT THIS ISSUE

Bruce Molnia, Ph.D. has studied Alaska's glaciers for more than 30 years. His first exposure was as a student on the Juneau Icefield in 1968. Since then he has taken tens of thousands of photographs of glaciers throughout the state and conducted research on coastal and fiord glaciers from south of Juneau to Cook Inlet. Molnia has authored several books and more than 100 articles, maps, and abstracts about Alaska's rivers of ice.

Many scientists and government researchers provided information on various aspects of glaciology. In particular, we thank: Tina Neal and Game McGimsey of Alaska Volcano Observatory/USGS, and Angela Roach, Department of Geological Sciences, Brown University, for help in updating the Aleutian Islands section of the text; Gary Prokosch of Alaska Department of Natural Resources, Ric Davidge of World Water, SA, and Al Schafer of Afognak Logging in Seward for their input on glacier ice harvesting; Dan Shain of Rutgers University for an update on ice worm research; Mike Fleming of USGS and Rachel Garcia of Mapping Sciences at the Bureau of Land Management for helping us understand the complexities of glacier mapping; and Dennis Trabant of USGS for answering a variety of questions.

EDITOR'S NOTE: Terms shown in color and boldface appear in the illustrated glossary on page 102.

Contents

Alaska's Rivers of Ice 4
   Introduction 5
   Water, Water — Everywhere! 7
*Present Day Distribution of
   Alaska Glaciers* 8
Tip of the Iceberg 10
   Anatomy of an Alaska Glacier System 13
Safe Passage 24
   Types of Alaska Glaciers 26
   Glacier Flow 29
   Glaciers and Climate 31
Glaciers and Alaska's Ice-Age Fossils 34
   Glacier Mass Balance:
     Accumulation versus Ablation 38
   Monitoring and Measuring
     Alaska Glaciers 39
Out of the Northern Ice 42
   Effects of Glaciers on Sea Level and
     Earth's Crust 45
   Exploration of Alaska Glaciers 47
   Life on Alaska Glaciers 54
Ice Worms 56
   Glacier Travel 57

Alaska's Glaciers: Area by Area 58
Where to See Alaska Glaciers 100
Glossary 102
Bibliography 109
Index 110

Alaska's Rivers of Ice

Introduction

Alaska exhibits many contrasts: majestic mountains, long, convoluted shorelines, and thousands of glaciers. Less than 200 square miles of glaciers exist in other states — Washington, Wyoming, Montana, Oregon, California, Colorado, Idaho, and Nevada — but all the glacier ice in the rest of the United States combined totals less than the area of a single large Alaska glacier. Alaska's glaciers range in size from tiny cirque glacier remnants covering fractions of a square mile to massive piedmont glaciers, such as Bering and Malaspina, each covering nearly 2,000 square miles, and each larger than the state of Rhode Island. With a glacier cover of about 29,000 square miles, Alaska has about one-half the glacier cover of Asia, about the same glacier cover as Russia, 2.5 times that of China and Tibet, 3 times that of South America, 6 times that of Iceland, 12 times that of Europe, 75 times the glacier cover of New Zealand, about 150 times the glacier cover of the rest of the United States, and about 1,000 times the glacier cover of Africa. Glaciers prevail as an important part of Alaska's landscape.

Most visitors to Alaska rarely see the largest of Alaska's glaciers, Bering and Malaspina, except perhaps from the window of a jet. Rather, the majority view a variety of medium to small valley glaciers — Mendenhall, Matanuska, Exit, Portage, or Worthington — accessible by automobile or bus, or they encounter the glaciers of Tracy Arm, Glacier Bay, Yakutat Bay, Prince William Sound, or Kenai Fjords from the deck of a tour boat, cruise ship, or ferry. Few Alaska visitors walk on a glacier or even touch glacier ice, save for those who collect the small ice chunks that drift to the shore of Portage Lake near Anchorage.

The few who are fortunate enough to explore the surface of a glacier or to linger in front of a tidewater glacier and watch it calve, form memories and take home stories that last a lifetime. This book presents the story of Alaska's glaciers to those who have not had the opportunity to examine firsthand the majesty, beauty, and power of a living Alaska river of ice.

No one knows the exact number of glaciers in Alaska, but it may exceed 100,000. Only about 650 of them have names. Since the late nineteenth century, more than 95 percent of Alaska glaciers that extend below elevations of 5,000 feet have been retreating, thinning, or stagnating. During the twentieth century, as many glaciers retreated and shrank in size, they separated into individual

FACING PAGE: *A nunatak juts from Grand Plateau Glacier in the Fairweather Range. The term originated with the Inuit people of Greenland. (R.E. Johnson)*

RIGHT: *Meltwater rushes from Shoup Glacier's tidewater terminus in Prince William Sound, flushing fine silt particles, rocks, and other debris from beneath the ice. (Nick Jans)*

Repeated **surges** *within Malaspina Glacier and its tributaries create abstract moraine designs on the glacier's massive piedmont lobe. Visitors flying between Anchorage and Juneau are treated to this bird's-eye view on a clear day. At the top left of this photo is Mount St. Elias, while at left center is massive Mount Logan. (R.E. Johnson)*

retreating tributary glaciers. Ironically, even though the area and volume of glacier ice in Alaska decreased, the number of glaciers increased. In summer 2000, I revisited Sherman Glacier, near Cordova. What had previously been a single, unnamed tributary that descended from the south-facing wall of Sherman Valley and connected to the main ice mass is now a **hanging glacier** and four separate cirque glaciers, none of which contact Sherman Glacier.

We are lucky to live in a period of time, both climatic and geologic, when glaciers exist and we have the technical capability to travel to them, fly over them, and photograph them. Less than 300 years ago much of the world remained ignorant of the existence of glaciers. Less than 200 years ago the majority of the geological community failed to recognize the importance of glacial erosion, and slightly more than a century ago, more than 99 percent of all Alaska glaciers lay secluded, completely unknown. Today, scientists are answering many questions about why glaciers form, how they move, and what they do. This book explains many of these phenomena.

gla·cier \ ˈglā-shər

1: a large, perennial accumulation of ice, snow, rock, sediment, and liquid water originating on land and moving down-slope under the influence of its own weight and gravity; many of the most spectacular glaciers on Earth occur in Alaska. — *Webster's Dictionary* (modified)
2: a dynamic river of ice.

Water, Water — Everywhere!

Water is a naturally occurring chemical compound made up of hydrogen and oxygen, H_2O, the most abundant chemical compound found on the surface of Earth. Water can exist in three physical states; as water vapor, a gas; as liquid water; and as ice, a solid. At different times, all three occur in the glacier environment. The three can and do frequently coexist. Not surprisingly, most of a glacier is composed of water, generally more than 95 percent.

Glacier ice constitutes the largest reservoir of fresh water on Earth, and aside from the oceans, the second largest reservoir of water. Most water resides in the oceans, 97.2 percent. The remaining 2.8 percent occurs in the atmosphere, lakes and rivers, ground water, soil, and glaciers. To look at this in a different way, if there were 1,000 drops of water on Earth, 972 would be in the oceans, less than 1 would be in the atmosphere, less than 1 would be in lakes and rivers, about 6 would be in ground water and soil moisture, and 21 would be in glaciers. Of the 2.8 percent that is fresh water, three-fourths is contained in glaciers. Three times more water exists frozen in glacier ice than all of the liquid fresh water on and in Earth.

Most glacier ice is located in Earth's polar regions, with 91.4 percent in Antarctic glaciers and ice shelves, and 7.9 percent in Greenland glaciers. The remaining less than 1 percent (0.7 percent) is found in the ice caps, icefields, and glaciers of North America, Asia, South America, Europe,

John Verhey shows off a chunk of melting iceberg calved from Shakes Glacier, 25 miles northeast of Wrangell in Southeast Alaska. Beginning in the 1970s, a handful of entrepreneurs tried commercially harvesting glacier ice but found that costs outweighed profits. (R.E. Johnson)

Africa, and the islands of New Zealand and New Guinea. Using the 1,000 drop analogy, 914 would be in Antarctica, 79 would be in Greenland, about 4 in North America, about 2 in Asia, and less than 1 in South America, Europe, Africa, and the islands of New Zealand and New Guinea combined. Of the 4 North American drops, 1 would be in Alaska.

Glaciers cover about 3.1 percent of Earth's surface, about 10.7 percent of the land, and about 5 percent of Alaska. Less than 20,000 years ago, during the peak of the most recent phase of the ice age, a period called the **Pleistocene**, glaciers veiled about 25 percent of Earth's land surface. If all glacier ice were to melt, sea level would rise 267 feet, flooding every coastal city on the planet. If only Alaska's glaciers melted, sea level would rise a little less than one foot.

Eroding lateral moraines, remains of Glacier Bay's retreating Muir Glacier, stick to glacially polished bedrock walls of Muir Inlet. (Bruce Molnia)

Present Day Distribution of Alaska Glaciers

Most Alaska glaciers lie within 100 miles of the Pacific Ocean. Today, glaciers span the entire southern perimeter of the state, from just north of the Canadian border in Southeast to more than half-way along the Aleutian Islands. Glaciers exist on the Alexander Archipelago, in the Coast Mountains, on peaks of the St. Elias Mountains and Alaska Range; in the Chugach, Kenai, and Wrangell Mountains; and in the Talkeetna Mountains, where most glaciers are rapidly shrinking. North of the Arctic Circle, more than 700 small cirque and valley glaciers exist in the higher summits of the Brooks Range, especially in the Romanzof and Franklin Mountains. Many volcanoes on the Aleutian Islands and the Alaska Peninsula are glacier-covered. As many as 50 small alpine glaciers remain on the backbone of Kodiak Island; more than 100 cirque glaciers persist in the Ahklun and Wood River Mountains south of the Kuskokwim River; and even two or three dot the Seward Peninsula's Kigluaik Mountains.

In all, counting every tiny cirque glacier, valley glacier, and every other permanent ice mass up to the size of Bering and Malaspina, Alaska may possess over 100,000 separate glaciers. Of these less than one percent have been named, and an even smaller percent have been investigated.

The less-than-precise measurement of glacial areas is due to at least three factors: first, in some remote areas no field observations have ever been conducted; second, in snow- and ash-covered areas it's difficult to ascertain where glaciers exist or where bedrock is buried by only a thin snow or ash cover; and third, many small glaciers, and even some larger ones, have completely disappeared since the time when last mapped.

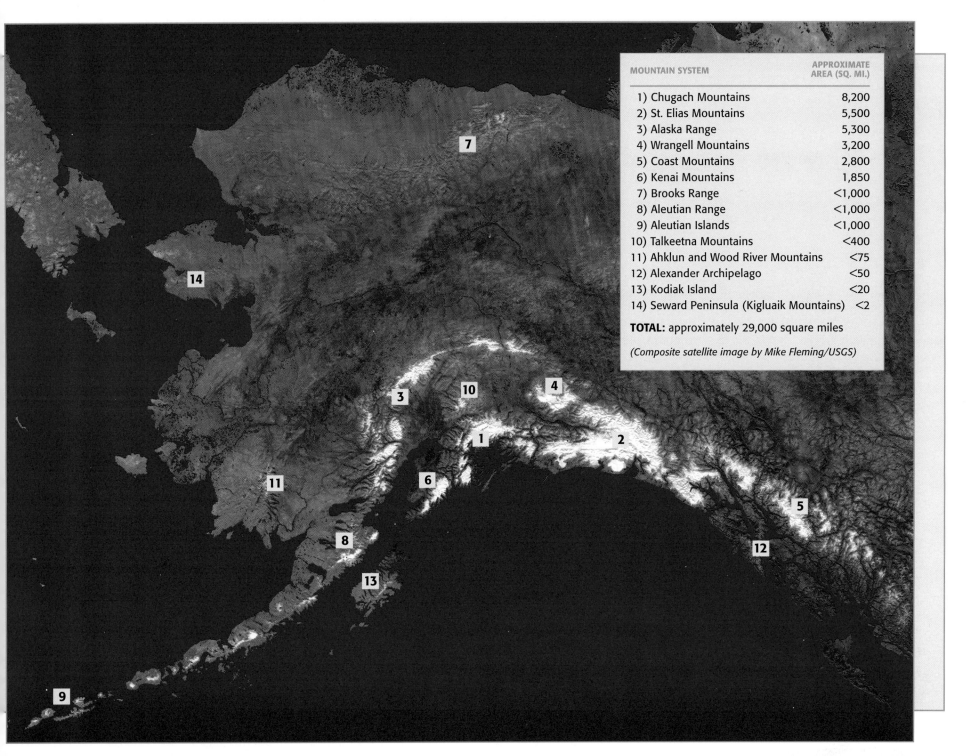

Tip of the Iceberg

By Susan Beeman, Associate Editor

ABOVE: *A few people tried harvesting glacier ice in Alaska, eager to sell it for profit, but most found the cost of gathering icebergs such as this one in Harris Bay, Kenai Fjords, an expensive venture. (Penny Rennick)*

Imagine dropping ice cubes into your drink only to hear them snap, pop, and crackle louder than any you've heard before. The sound is called ice sizzle or bergy seltzer. One cube explodes, liberating ancient air trapped inside and displacing liquid over the lip of the glass. These are no ordinary ice cubes. These are from glacier ice, under intense pressure — up to 1,000 pounds per square inch (psi). Compare that to the 30 or 40 psi needed to properly inflate a car tire and you'll have an idea of the power sealed inside glacial ice, ice that has been compressed for hundreds or thousands of years.

In the late 1970s, a few entrepreneurs began harvesting Alaska icebergs that calved into the sea from tidewater glaciers in Southeast with the express purpose of selling the ice as a novelty item. This was the routine: hoist the huge bergs onto a barge, transport them back to a dock, rinse off the salt water, cut them with a power saw, and load them into refrigeration vans destined for various markets around the world.

The Japanese market was the most lucrative, though for a time, some Hawaiian hotels bought the special ice too. Al Schafer of Afognak Logging in Seward remembers when he'd take his barge and tugboat out to Aialik Bay, on the Gulf Coast of the Kenai Peninsula, and haul icebergs on deck with a crane. "You needed that big heavy equipment," he says, "you couldn't just go out and pick up icebergs in a skiff." Schafer says the ice he sold was shipped to Los Angeles, then sent to Tokyo, remaining the entire time in the original refrigeration van. The Japanese were drawn to anything old, Schafer explains, and the novelty of glacier ice from overseas was a hit. Wetco, an Anchorage-based company involved in bottling water and water-quality issues, marketed Alaska's glacier ice to Japan at about $3.50 per pound.

Only a few people ventured into the business. Enthusiasm for it has dwindled in recent years due to the high expenses and logistical difficulties of ice transport. The last reported wholesale price for glacier ice, according to the Alaska Department of Natural Resources (DNR), was $500 per ton. Gary Prokosch, chief of the Water Resources Division of DNR, said his office gets only two or three requests for information annually now, and that no active harvesting permits exist today.

Alaska's glacier ice is considered a resource, and as such, the government regulates its harvest. Potential harvesters of significant amounts (defined by DNR in terms of tons and environmental effects) must apply for a permit, pay a $500 filing fee, and pay for a legal advertisement in at least one issue of a local newspaper in the area near the proposed ice harvest.

Besides being a source of ice, glaciers store 75 percent of Alaska's fresh water. As drinking water is a dwindling commodity in the desert Southwest and other parts of the world, the state's

BELOW: *Tourists fill their coolers with chunks of ice that have floated to the edge of Portage Lake 40 miles southeast of Anchorage. Though Portage Glacier, once visible from Begich, Boggs Visitor Center, has retreated around a valley wall out of sight, icebergs still wash ashore. (Roy M. Corral)*

glacial meltwater has received national and international attention over the years. Several schemes to transport fresh water from Alaska to the Lower 48 have been considered. Former governor of Alaska Walter J. Hickel proposed an undersea pipeline that would deliver water from the mouth of the glacially fed Copper or Stikine Rivers in Alaska to Lake Shasta in northern California, but for various reasons the plan was never implemented.

Others take Alaska's supply of fresh water seriously, too. In 1977, Prince Mohammed al Faisal of Saudi Arabia spent an estimated $1 million gathering information on alternative water sources for his parched country. His research included conferences attended by engineers and scientists from around the world, and at one such gathering, Prince Faisal had an iceberg helicoptered, ferried, flown, and trucked all the way from Alaska to Ames, Iowa for $14,000 just to test his ice transport hypothesis. When it arrived, he had pieces of the crackling ice served in drinks to his guests.

A 1994 report on water exports by Ric Davidge, former director of water and chief of the Alaska Hydrologic Survey, DNR, says, "The Worldwatch Institute, in July 1993, reported that no less than 26 countries have larger populations than their water supplies can adequately support. ... Global water use has more than tripled since 1950." Davidge concludes, "Importing Alaska water to the southwestern states and Mexico is economically, environmentally, and politically far more attractive than the alternatives available, especially desalinization or new surface storage and transport."

It's a good question: Should we export Alaska's glacier ice and water? From multimillion-dollar projects to a few sizzling cubes, it's one that likely will be debated for years to come. What might on the surface seem to be improbable may only be the tip of the iceberg.

Anatomy of an Alaska Glacier System

In simplest terms, an Alaska glacier is a mixture of ice and rock that moves downhill over a bed of solid rock or sediment under the influence of gravity. But a glacier contains more than just moving ice and rock. At different times of the year, liquid water may be on, in, or under the glacier. As this water flows, it transports sediment. As the ice moves, it modifies the bed over which it flows, sometimes eroding and sometimes depositing sediment. In reality, an Alaska glacier is a complex and dynamic system that is composed of the glacier, its bed and surrounding environment, and the sediment it erodes and deposits. All are interrelated. All are continuously changing in response to short-term fluctuations in temperature and precipitation and to longer-term fluctuations in climate.

To understand how an Alaska glacier system behaves, let's examine the components of a glacier and its environment: first, the glacier ice; second, the complex array of deposits a glacier produces as it advances, melts in place, or retreats; and third, the bed, valley, fiord, or channel in and over which the glacier flows.

FACING PAGE: *Stranded as the result of a* jokulhlaup, *or glacial outburst flood, grounded icebergs melt in a valley below Nelchina Glacier in the Chugach Mountains after an ice-dammed lake gave way. (Fred Hirschmann)*

RIGHT: *Michaels Sword, a 6,840-foot peak on the Juneau Icefield one mile from the Alaska-British Columbia border, is a good example of a glacially carved* horn. *(John Hyde)*

GLACIER ICE

Glacier ice is different from all other ice on Earth. Unlike lake ice or sea ice or even refrigerator ice, which form by the freezing of liquid water, glacier ice evolves though the metamorphism of snow. Six-sided, elongated crystals form through compaction, compression, and recrystalization of individual snowflakes to eventually produce glacier ice. This metamorphism happens not over night, but rather over a number of years. As each successive year's snowfall creates a new layer on top of the older layers, weight and pressure increase, and transformation takes place. The more new snow that accumulates, the greater the overburden weight and pressure, and the quicker the change occurs. Each new layer alters the density, volume, and crystal structure of the snow below.

In simple terms, density is the comparison of the weight of any object to an equal volume of liquid water. By definition, a cube of liquid water that measures one centimeter by one centimeter by one centimeter in size

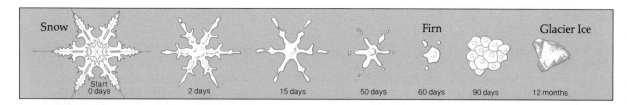

LEFT, TOP: *Delicate snowflake points disappear as melting and evaporation occur. The resulting water vapor condenses near the center of the crystal, making it smaller, rounder, and denser. Technically, glacier ice is considered metamorphic rock, though its melting point is much lower than that of other rocks. (ALASKA GEOGRAPHIC® illustration, modified from U.S. Army Corps of Engineers)*

LEFT, BOTTOM: *A glacial stream channel, once full of water, now harbors only a trickle. (Bruce Molnia)*

(one cubic centimeter) has a weight of one gram and a density of one. Newly fallen snow consists of individual six-sided crystals that have a density of 0.1 to 0.3. Glacier ice has a density of about 0.9. Thus, in the metamorphic process that changes snow to ice, a volume decrease of up to nine times occurs. Snow, generally after surviving one summer melt season, converts to a material called firn. In each successive year, the firn becomes denser as air is forced out. The density of firn ranges from about 0.4 to 0.8. As firnification (the process of changing snow to firn to glacier ice) occurs, original layering disappears and individual snowflakes merge to form granules of solid ice. These original layers frequently undergo deformation and contortion during firnification. The change from snow to glacier ice may take up to 20 years in areas of high accumulation or many decades in areas of low accumulation. Warming above the freezing point may completely destroy crystal formation and establish ice layers instead.

Finally, when an interconnected network of ice crystals forms that will not permit the movement of air and liquid water, glacier ice is created. The thickness of Alaska glacier ice accumulations varies from a minimum of a few hundred feet to more than 3,000 feet. For ice to flow, it needs to be more than 50 feet thick because it flows in response to overburden pressure (the weight of the material above) and gravity. Because of differences in glacier motion and flow, ice crystals of different size and type are segregated into bands or layers called folia with the layering or banding termed foliation. Often, sequences of folia consist of alternating layers of clear and bubbly ice, or alternating layers of fine-grained and coarse-grained ice. Rarely does the foliation correspond to the original snow and firn accumulation layers.

Glacier ice has a density range of 0.88 to 0.92, depending on the amount of air that is trapped between and within crystals, and often appears blue. It is impermeable to air and water. Individual ice crystals may grow to more than a foot long. By comparison, refrigerator ice, sea ice, and lake ice are not crystalline, are much less dense, and contain much more air. All form quickly without any overburden pressure. Glacier ice looks blue because the physical characteristics of water molecules absorb all colors except blue, which is reflected. At one time, glaciologists used the French term *névé* interchangeably with firn, but in today's vocabulary, *névé* means an area covered with perennial snow, or the area of accumulation at the upper end of a glacier.

Air trapped within glacier ice is frequently subject to pressures that may exceed 1,000 pounds per square inch. When glacier ice calves at near-sea-level-pressures, producing icebergs, bubbles contained in the icebergs are in equilibrium with the depths and pressures under which they formed. As melting progresses and the surrounding glacier ice barrier thins, the pressurized air bubbles escape, frequently producing a resounding pop. With a sound known as ice sizzle or bergy seltzer, hundreds of these bubbles break almost simultaneously, creating a continuous crackling sound reminiscent of Rice Krispies.

The energy released through ice sizzle is intense. Sound generated by ice sizzle has been detected up to 100 miles by underwater hydrophones, sound detection devices used in tracking distant ship movements.

***Rock glaciers**, such as these spilling from **cirques** in the rugged Alaska Range east of Cantwell, move much more slowly than do ice glaciers. Ice forms between the rocks, but is not usually visible on the glacier's surface. (Bob Butterfield)*

Glaciologists laughingly tell of celebrating the completion of a deep ice core drilled to more than 2,000 feet. The crew chilled the party's liquid refreshments with ice from the bottom of the core. The celebration quickly turned disastrous as glass after glass cracked under the impact of the bursting high-pressure bubbles.

From a vantage point looking down-valley, wave **ogives** *usually appear as arcs between medial moraines on Vaughn Lewis Glacier, in the Coast Mountains. Ogives form at the base of* **icefalls**; *these sets of ogives support small, turquoise, glacial lakes. (Bruce Molnia)*

To experience ice sizzle on a greatly reduced scale, a person needs only a glass of water and a refrigerator ice cube. Although there is no pressure difference between the water or the bubbles in the ice cube, small audible pops can be heard as the cube melts. As melting continues, cracks may develop along bubble planes and the ice may split or fracture. In a similar fashion, glacier icebergs may fracture or roll over following intense episodes of ice sizzle. Even beached icebergs can be heard "popping away" as they sit on the sand. If you ever have the opportunity to be near an iceberg, listen closely and you will hear the symphony of bubbles as they play the themes of ice sizzle and bergy seltzer.

Liquid water constitutes an integral part of an Alaska glacier. Streams exist in, on, and under most of them. During summer months, much of a glacier's surface becomes wet with streams flowing into **crevasses** or narrow tubular chutes called **moulins**, or glacier mills. Running water frequently melts channels 15 feet wide or more into the surface of a glacier.

GLACIER SEDIMENTS AND DEPOSITS

Sediment of any size, whether microscopic silt particles or boulders larger than a house, is called **drift**. Drift can be deposited on the surface of a glacier in several ways: by running water, by avalanches, or through the air as falls of rock, dust, or volcanic ash. Drift can also accumulate around the entire perimeter of the bed of a glacier and can build up distinct layers within the glacier. The term stems from the early impression that all these sedimentary deposits attributed to glaciers were the result of the great Biblical Flood described in Genesis 5:28-10:32.

Sediment deposited directly by ice is called **till**, and is generally poorly sorted and not layered. Glacial geologists recognize two types of till: lodgement till, material plastered in place by ice as it moves forward, and **ablation** till, material dropped to the ground as stagnant ice melts in place. After a glacier has melted, a blanket of till, called ground moraine, covers all surfaces over which a glacier flowed.

If the terminus of a glacier remains essentially in the same place for a period of time, a ridge or mound of till develops adjacent to the ice. If this moraine is the farthest down-valley, marking the maximum extent of the glacier, it is called the terminal or end moraine. Up-valley are successive

moraines, each representing a position where the ice margin temporarily stood still during a period of retreat. These are termed recessional moraines. Both terminal and recessional moraines often block meltwater streams, creating moraine-dammed glacial lakes. *Jokulhlaups* are outburst floods that form through the failure of glacier-ice-dammed or moraine-dammed lakes, or through subglacial volcanic eruptions. The

Dean Rand paddles his kayak around calved icebergs in Prince William Sound's Nellie Juan Lagoon. Boating near icebergs can be risky, since they roll or break apart as they melt and their center of gravity shifts. (Patrick J. Endres)

Crevasses dwarf a Cessna 206 as it flies sightseers over Hubbard Glacier. The glacier was named by USGS geologist Israel Cook Russell in 1890 for Gardiner G. Hubbard, regent of the Smithsonian Institution and founder and first president of the National Geographic Society. (R.E. Johnson)

term is adopted from Iceland, where a detailed recorded history of periodic glacier floods extends back to the fourteenth century. Sedimentary deposits document many Alaska glacier *jokulhlaup* events.

As a glacier slides over bedrock, it carries and drags a load of sediment and rock known as the basal till layer. This layer, frozen into the basal ice, polishes and scrapes away small rock particles, known as **rock flour**, through abrasion.

In addition to abrasion, which produces

material on the order of fractions of an inch to inches, glaciers quarry large blocks of bedrock prepared for transport by the freezing and thawing of water in cracks, joints, and fractures. This transportation of larger blocks is known as glacial plucking. Plucked particles may reach dimensions of tens of feet or more.

Two other types of moraines occur on the surface of a glacier: lateral moraines and medial moraines. A lateral moraine develops on each side of a valley glacier and consists of abraded sediment and plucked rock material from the valley walls or rock and sediment that avalanche onto the surface of the ice. Where two valley glaciers coalesce, lateral moraines merge to form a medial moraine, one that is now in the middle of the combined glacier. Lateral moraines are frequently preserved after a glacier melts away, generally as a veneer of sediment plastered on the valley wall. Medial moraines rarely survive, as meltwater streams rework them and carry away much, if not all, of the sediment. Large valley glaciers or piedmont glaciers may show 20 or more distinct medial moraines indicating the enormous number of tributary glaciers that have joined to form the large glacier system.

Running glacial meltwater transports much glacially eroded sediment, depositing it in layers in front of the glacier on a broad, low-angle surface known as an outwash plain. Occasionally, a block of ice is left behind during retreat or carried by a meltwater stream onto the outwash plain, then buried in the sediment. As the ice melts, a depression, called a kettle, forms, and continues to enlarge until all the ice melts. A pitted outwash plain is one with many kettles.

On the outwash plain, the volume of sediment deposited is often so great that it cannot be transported at one time by the quantity of water available. Braided streams, dominated by meandering channels and ever-changing bars, continuously rework the sediment. Examples are the Copper River, north of Cordova; the Alsek River, south of Yakutat; and the Susitna River, south of the Alaska Range. These braided streams transport rock flour and deposit it in large plumes in lakes, bays, Cook Inlet, and the Gulf of Alaska. Valleys leading away from the front of a glacier also may receive a substantial amount of outwash fill. These are termed valley train deposits.

Some sedimentary deposits accumulate where running water comes into direct contact with ice. The most common type of deposit is a poorly sorted sand and gravel mass, called a kame, which forms in direct contact with stagnant ice. Kames, which generally have some stratification, develop within cracks, holes, or crevasses in the ice or between the ice and the land surface. A well-stratified deposit known as a kame

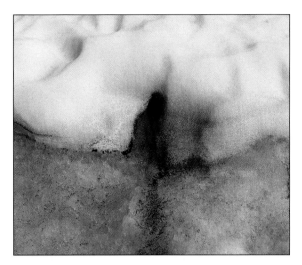

ABOVE, LEFT: *Red-pigmented green algae grows along a meltwater stream, turning the snow pinkish. Red algae feeds itself through photosynthesis and can spread rapidly across a glacier's surface during summer. (Bruce Molnia)*

ABOVE: *Like other metamorphic rocks, glacier ice deforms, producing faults and folds in response to stresses and movements. (Bruce Molnia)*

terrace often builds up between the glacier and its valley wall. When the ice melts, the terrace remains along the valley side and is often confused with a lateral moraine. Exam-

Hundreds of visitors come to Alaska every summer to experience glaciers firsthand. Here, Brian Heinselman listens to groans and creaks erupting from Meares Glacier, 40 miles west of Valdez. Meares is one of few glaciers in the state that is advancing. (Patrick J. Endres)

ination of the feature's structure, though, can help differentiate between them. Since a kame is well stratified, many distinctive individual beds and groups of beds are apparent. This is typical of water-deposited material. A lateral moraine almost never has any bedding. It forms from unsorted material being dumped on the edge of the glacier. The shape may be the same, but the internal structure is completely different.

Crevasse fills and **eskers** are other common kinds of kame deposits. Running water on the surface of a glacier often washes sediment into crevasses where the sediment remains. When the ice melts, the crevasse fills persist as long, steep, narrow, stratified ridges. Running water under a glacier may erode a meandering channel in the ice that can fill with sediment. As the glacier melts, a meandering, stratified, sediment ridge called an esker may emerge from underneath the retreating terminus. Eskers vary in height from several feet to more than 100 feet, and in length from a few hundred feet to tens of miles.

Where glacier meltwater streams empty into lakes or bays, deltas form. When the glacier terminus ends in a body of water, blocks of ice break off and float away. The process of the blocks separating from the main glacier is known as calving. Any sediment embedded in the icebergs that drift away is referred to as being ice-rafted away from the glacier. As the iceberg melts or turns over, the sediment it carries falls to the bottom of the lake, bay, fiord, or ocean. During the Pleistocene, such ice-rafted sediment was deposited in the Pacific Ocean. Rocks which originated in the St. Elias and Chugach Mountains have been recovered in cores collected from the sea floor more than 1,000 miles south of their Alaska source, dumped there by drifting icebergs.

GLACIER VALLEYS, FIORDS, AND BEDROCK FEATURES

Glaciers generally flow along the path of least resistance. They commonly occupy a stream valley or a fault trench and widen and modify it by abrasion and plucking. Streams flow in V-shaped valleys. Glaciers modify the shape of these stream valleys and change them to a characteristic **U-shaped** glacial valley. In the process of widening and deepening coastal valleys, glaciers often erode the valley floor to below sea level and extend the valley onto the continental shelf or into deep bays, where valleys become **fiords**. Alaska has hundreds of fiords, the best known located along the Gulf of Alaska coast, in the Inside Passage, Glacier Bay, and Prince William Sound. Large fiords may be up to five miles wide and have nearly two miles of vertical relief.

Glacier valleys are not only rounded or U-shaped in cross section but have rounded amphitheater-like basins cut in their upper ends or sides. These half-bowl-shaped depressions, called cirques, have lips at their lower ends, called thresholds. Small glaciers within cirques may have tongues of ice cascading or even avalanching down the valley walls until they reach the main glacier. Glaciers change topography. They tend to round, deepen, oversteepen, and streamline the surfaces they contact. Glaciers may be the most efficient mechanism on Earth for erosion.

Valley glaciers often develop in parallel, closely spaced valleys. As individual glaciers erode and widen their valleys, the bedrock ridge that separates two adjacent glaciers narrows and becomes oversteepened. A series of narrow, jagged spires, much like the edge of the blade of a serrated knife, develops along the ridge crest. These features are called *arêtes*, after the French word for fish

BELOW: *Among the life forms that thrive in glacial environments, harbor seals use icebergs for birthing pups in the spring. These were photographed in Tracy Arm-Fords Terror Wilderness, south of Juneau. (John Hyde)*

RIGHT: *Resembling choppy swells on a stormy sea, Matanuska Glacier crevasses form as ice creeps over steep terrain. The glacier, visible from the Glenn Highway, is accessible by a side road 102 miles northeast of Anchorage. (Fred Hirschmann)*

BELOW: *Like frosting on a cake, new snow blankets huge ice blocks at Mendenhall Glacier's terminus, and locals take advantage of a sunny afternoon to visit this giant in their back yard. (John Hyde)*

RIGHT: *Margerie Glacier, which originates on the flanks of Mount Fairweather, snakes its way down-valley to end in Tarr Inlet, in Glacier Bay National Park. Tidewater glaciers such as this terminate in a fiord or the ocean and actively calve icebergs into the sea. (Harry M. Walker)*

bones. Eventually, with continued erosion, the *arête* disappears and the glaciers merge.

As glaciers erode headward, or up-valley, into the side of a mountain, they modify the configuration of the mountain's summit to form a steep-sided, sharp-pointed, pyramidal mountain peak called a horn. When four glaciers erode a symmetrically shaped horn, it is termed a matterhorn after the famous peak in the Swiss Alps. Horns, *arêtes*, and cirques, the most common bedrock features that can be observed in recently deglaciated parts of Alaska, are often seen poking through the edges of Alaska's many valley glaciers.

As a glacially eroded area emerges from melting ice, many large, rounded, asymmetrical bedrock knobs, called **roches moutonnées**, (French for fleecy rocks) begin to emerge. These knobs, which form subglacially, have a gentle slope on their up-glacier side and a steep to almost vertical face on their down-glacier side. This geometry is due to the glacier gliding and gently over-riding the near side and then plucking blocks of rock from the knob's far side as it flows past. Glacial geologists use the shape and orientation of *roches moutonnées* to help interpret glacier flow directions.

When a glacier thins, as melting increases or as a recession begins, a sharp line that marks the maximum extent of the glacier's margins appears. This line, which may either be a change in type or presence of vegetation or a change from weathered to unweathered bedrock is called a **trimline**. Comparison of the height of the most recent trimline above a glacier with the height of the ice surface is useful in determining a glacier's health. Where multiple trimlines occur, comparison of the age of vegetation associated with each trimline may provide clues to the timing of significant recent changes in an individual glacier.

As glaciers slide over their beds they often polish, fracture, groove, or striate the surface. When ice motion is irregular, large particles may grind along the bottom and carve individual or multiple crescent-shaped fractures, gouges, or chatter marks, generally perpendicular to the direction of glacier flow.

Ablation moraine and new plant growth cover the lower end of Muldrow Glacier. Upper reaches of Muldrow have long been an access corridor for climbers of North America's highest peak, Mount McKinley, shown here. (Hugh S. Rose)

Striations and grooves are long, straight, parallel furrows, oriented in the direction of glacier motion.

Safe Passage

By Ed Darack

EDITOR'S NOTE: *Ed Darack is the author of* 6194 Denali Solo; Wind-Water-Sun; *and a new book,* Wild Winds: My Adventures in The Highest Andes, *which chronicles his climbs of the highest mountains in South America, including Aconcagua, Ojos del Salado, Cerro Pissis, Llullaillaco, and Nevado Sajama. His articles and photos have appeared in previous issues of ALASKA GEOGRAPHIC® including* Fairbanks, Climbing Alaska, *and* World Heritage Wilderness: From the Wrangells to Glacier Bay.

Listen to the moans of a shifting crevasse field, gaze at sinuous medial moraines, feel a rush of chilled air descend an ice-locked valley, peer back in time at cerulean layers of compressed snowstorms — an infinite array of adventures awaits those who set foot, snowshoe, or ski on one of the many glaciers of the North. There is more to glacier travel, however, than meets the eye.

Prospective glacier travelers need to be wary of the dangers associated with these parts of the mountain world. First and foremost: Crevasses. A glacier is a river of ice — a river that flows very slowly. And it only *flows* below a depth of about 150 feet under the surface, where the pressure is great enough to allow ice to deform plastically. The upper 150 feet of a glacier is known as the zone of fracture, particularly in regions where glaciers merge with one another or change slope. The large cracks that form (often called "slots" by climbers) are termed crevasses and can be wider than a school bus is long.

Crevasse fields are treacherous enough when you can see individual slots; however, the real danger lies with a blanket of snow — from a recent snowstorm or remaining from a past winter — that hides crevasses. Snow bridges, tenuous spans that often sag and become slightly discolored during summer months, reveal the general locations of individual crevasses, but make it impossible for glacier travelers to know how much weight they can hold. By the end of a warm summer, what was once a smooth glacial surface is a hodgepodge of shattered ice towers. Navigating these labyrinthine sections — even knowing where these sections start and end — is something that any experienced mountaineer will tell you comes only with time, and is more feel than anything else. Heavily traveled mountaineering

Jamming skis firmly in snow on Ruth Glacier, a climber demonstrates how to set a rope anchor for crevasse rescue. Practicing rescue techniques on a regular basis helps assure group members' safety. (Harry M. Walker)

routes, such as the West Buttress of Denali, often have wands (thin bamboo sticks with fluorescent orange flags) placed by previous climbing teams, marking each side of a snow bridge. Nevertheless, no one should ever venture onto a crevassed glacier assuming there is a safe route explored and marked. Safe passages can change by the day, as snow bridges collapse and crevasse walls topple over.

The best way to safely explore a glacier is to rope up with at least one other person, and to wear snowshoes or skis, which distribute your weight over a greater area than your feet alone provide. The ideal rope team is a group of three, as this balances speed and efficiency of movement with safety in numbers. Experienced rope teams keep the rope taut between teammates, with the leader carefully looking, listening, and probing with ski poles or a special type of probe pole that consists of two ski poles attached end-to-end. If a member of the rope team pops through a snow bridge, then the other two arrest the fall, tie off the rope, and the fallen member either climbs to safety with mechanical ascenders, or the two on the surface hoist him out. A "Z-system," using pullies and carabiners, in which the rope is doubled back forming a "Z" pattern is one method of crevasse rescue. This is not easily done, however. Imagine falling into a crevasse, hanging upside down with an 80-pound pack, wedged between converging walls of ice that average minus 40 degrees Fahrenheit. Quite a bit of energy is needed to escape a situation like that. Now imagine that you are unconscious with a compound fracture. Although it may look easy from afar, working out a good rope team takes a lot of practice and patience, but it's essential for safe travel.

Other considerations for glacier travel fall within the scope of general mountain safety and proper preparedness for bad weather — adequate clothing, ample food and water, emergency shelter, and communication. While many of these preparations assume the weather to be cold and inclement, never forget the possible dangers of the opposite. In late June of 1990, while I was the base camp manager on the southeast fork of Kahiltna Glacier near Mount McKinley, I skied down to the main Kahiltna Glacier and was overcome by the heat. I had a small thermometer with me and to my disbelief I read the temperature was approaching 100 degrees Fahrenheit! The crystal-clear air and high sun combined with an exposed dark layer of volcanic ash from the 1989 Mount Redoubt eruption created that condition. But once the sun fell behind Mount Crosson that night, telltale crackling sounds of refreezing surrounded me. The temperature dropped to 20 degrees Fahrenheit — a loss of 80 degrees within 12 hours.

One final note on good glacier weather — travelers need adequate protection from the sun. While clear, warm conditions make it comfortable to amble about a glacier with just a T-shirt and shorts, you can never have enough sun protection — even on the bottom of your chin. The sun's rays come at you from every direction on a glacier; sunscreen, a good sun hat, and glacier glasses with side guards are imperative. Don't tread ice without them.

Roped up and decked out with all the necessary climbing gear — helmet, ice axe, carabiners, and crampons — Bill Mohrwinkel jumps a crevasse on Matanuska Glacier. (Tom Bol)

Types of Alaska Glaciers

Glaciers are classified either by their size, from tiny cirque glaciers to continent-covering ice sheets; by their thermal characteristics, polar versus temperate; or by other traits, such as ending in tidewater. The classification used here is based on size, shape, and geographic location. Glaciers in Alaska range in size from smaller than a football field to larger than the state of Rhode Island.

Hanging glaciers cling to 3,000-foot peaks above Laughton Glacier in the Sawtooth Range a few miles northeast of Skagway. Unsorted till litters the area, transported by ice and deposited at its current position. (Bob Butterfield)

A cirque glacier is a small glacier that forms within a cirque basin. Today, many Alaska glaciers are not thick enough to reach the lip and they remain within the basin. Others overtop the threshold and flow downslope. Cirque glaciers can be found on the summits of many of the highest mountains in the state, including Mounts McKinley, St. Elias, and Fairweather. Small cirque glaciers are sometimes called glacierets.

Valley glaciers originate from ice accumulating in one or more basins or cirques, or from overflows from an icefield or ice cap on top of a plateau. Many of the larger valley glaciers in Alaska exceed 20 miles in length. Hubbard Glacier, more than 70 miles, is the longest.

The most spectacular examples of piedmont glaciers in Alaska are Malaspina and Bering Glaciers, each with piedmont lobes covering areas greater than 800 square miles. A piedmont glacier forms when one or more valley glaciers flows from a confined valley onto a plain where it can expand into a broad, fan-shaped ice mass at the base of the mountains.

An icefield covers a mountainous area where large interconnecting valley glaciers are separated by mountain peaks and ridges, which project through the ice as nunataks. The lower parts of the valley glaciers serve as outlet glaciers and drain ice from the icefield. Alaska icefields include the Stikine, Juneau, Harding, and Sargent; each has an area of more than 500 square miles.

An ice cap is a dome-shaped or platelike cover of perennial snow and glacier ice that completely covers the summits of a mountain mass so that no peaks emerge through it. The term also applies to a continuous cover of snow and ice on an Arctic or Antarctic land mass that spreads outward in all directions because of its mass. Ice caps have areas of less than about 20,000 square miles. Ice caps cover several Canadian Arctic islands. During the Pleistocene, parts of southern and southeastern Alaska were covered by large, ice-cap-like subcontinental glaciers.

Ice sheets or continental glaciers are vast accumulations of glacier ice and snow that completely blanket a large land mass. The Antarctic and Greenland Ice Sheets are the largest on Earth. The Antarctic Ice Sheet covers more than five million square miles and in places exceeds 14,000 feet in thickness.

An ice shelf is a floating glacier that forms when a land-based glacier extends into the ocean. Antarctica has many of these floating

In undulating waves, Bering Glacier's immense piedmont lobe slips toward the Gulf of Alaska coast, calving icebergs into several lakes and rivers that carry the resulting meltwater out to sea. The Chugach Mountains rise in the background. (Bruce Molnia)

glaciers, with the largest, the Ross Ice Shelf, covering six times the area of Alaska's glaciers.

Thermal characteristics of glaciers fall into two categories: temperate and polar. Temperate glaciers, including most in Alaska, remain warm enough during a part of the year for liquid water to form through melting. Polar glaciers evolve where the annual temperature stays below the freezing point so that liquid water is never present. In Alaska, polar glaciers are confined to high elevations.

Several other terms describe unusual occurrences of glacier ice, such as rock glaciers, tidewater glaciers, and **reconstituted glaciers.**

Rock glaciers frequently head in a cirque and consist of a valley-filling accumulation of angular rock blocks. They resemble a glacier in shape, but have little or no visible ice at the surface. Investigations have shown that ice fills the spaces between rock blocks and that rock glaciers move, although very slowly.

A reconstituted glacier, also known as a reconstructed glacier or a glacier remanié, forms when pressure melting (**regelation**) joins ice blocks that accumulate below the terminus of a hanging or cirque glacier. Much of the ice in a reconstituted glacier accumulates through avalanching. Once reconstituted, these glaciers behave like normal valley glaciers.

Hikers slake their thirst on Norris Glacier, part of the Juneau Icefield. Though rarely found in high glacial streams, the parasite Giardia lambia *can infect humans who ingest it. Even water that looks pristine can carry this protozoan. (John Hyde)*

RIGHT: *The wall of Harvard Arm, in College Fiord, displays marks left by a Little Ice Age advance of Harvard Glacier. Rocks and other particles carried by moving ice cut chatter marks, gouges, and striations as they flow over the glacier's bed and walls. (Bruce Molnia)*

BOTTOM RIGHT: *A glacier resembles the outer layers of Earth (lithosphere and asthenosphere) in that its surface zone is brittle and cracks, while the layer underneath is flexible. Note that crevasses do not extend below the zone of brittle flow. (ALASKA GEOGRAPHIC® illustration)*

Glacier Flow

Flow within an Alaska glacier is a function of ice thickness, geometry, depth of the channel or valley, and temperature. In simplistic terms, the upper 100 to 150 feet of a glacier deforms and flows in a brittle fashion, often developing elongate cracks that may extend hundreds or thousands of feet across the glacier's surface. The cracks, which often change in size and shape as a glacier flows down-valley, are called crevasses. Below that depth, pressure on the ice increases, flow is more plastic, and the glacier moves fastest here. It also slides the fastest here in its center. Toward the bottom of the glacier, friction with the bed decreases flow rate. At the head of a valley glacier, a single large crevasse or series of smaller crevasses develops where moving ice pulls away from the rock or cirque wall. This crevasse system is called a *bergschrund*.

Typically, glaciers flow at rates of inches to two to three feet per day. Some glaciers, however, occasionally experience sudden, large-scale, short-lived increases in their rates of movements, 10 to 100 or more times faster than normal. These rapid rates of movement are called surges. Austin Post, a retired U.S. Geological Survey (USGS) glaciologist who has studied surges for nearly 50 years, attributes these rapid movements to a remarkable instability that occurs at periodic intervals in certain glaciers. He suggests possible causes may be related to bedrock roughness or permeability, anomalously high groundwater temperature, and abnormal geothermal heat flow. Others have suggested that certain surges were the result of earthquakes, avalanches, and local increases in snow accumulation. However, recent studies of Variegated and Bering Glaciers suggest that blockage of subglacial drainage channels causes an increase in water pressure and volume below and within the glacier. This

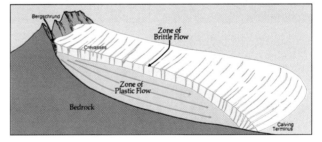

in turn results in a thin water layer at the base of the glacier that lubricates the bed and causes the glacier to move like a hydro-planing car.

Based on aerial photographs, Post has identified more than 200 North American glaciers that either were currently surging, or that had surged in the recent past. Alaska hosts at least two-thirds of these surging

ABOVE: *Multiyear stresses carve deep crevasses into Bering Glacier. The upper 100 to 150 feet of a glacier's surface is called the zone of brittle flow.* (Bruce Molnia)

ABOVE, RIGHT: *Taken from an altitude of 8,000 feet, an aerial view of Malaspina Glacier's contorted surface yields alternating bands of ice and moraine debris.* (R.E. Johnson)

glaciers. Post identified the glaciers on the basis of intense crevassing and folded surface moraines that were produced by rapid ice displacements, characteristic of surge movements, or distinctive surface features that resulted from previous surges. He found that surging glaciers have specific geographic occurrences. In Alaska, they are restricted to the Alaska Range, eastern Wrangell Mountains, eastern Chugach Mountains, and the St. Elias Mountains near Yakutat and Glacier Bay. No surging glaciers have been identified in the Coast Mountains, west and central Wrangell Mountains, west and central Chugach Mountains, Kenai Mountains, or the Brooks Range. Post's analysis of surging glaciers showed that they exist in maritime to continental climates and in temperate to subpolar environments. Surging is independent of elevation, bedrock type, valley configuration, glacier orientation, or size. Malaspina and Bering Glaciers show evidence of multiple surges. During a short interval of its 1956-1957 surge, Muldrow Glacier sped forward as much as 1,150 feet per day. During its 1993-1995 surge, Bering Glacier's terminus advanced more than 330 feet per day. Variegated Glacier, in Russell Fiord, and Black Rapids Glacier, in the Alaska Range, are being intensely investigated to unravel the origin of surges.

Ogives are arcuate bands or undulations in the ice at the surface of a glacier that occur in patterns generally oriented in a convex down-glacier position. Two types occur: band ogives, alternating light and dark bands on a flat, smooth, glacier surface; and wave ogives, undulations of varying height in the surface of the ice. Ogives form below icefalls, areas where glaciers cascade over steep bedrock slopes. The banding results from differences in flow rate in summer versus winter. One pair of dark and light bands or one pair of large and small bands represents the total flow of a full year.

Glaciers and Climate

Complex, dynamic climate cycles govern the planet. The geologic record indicates that glaciers have waxed and waned on Earth for more than a billion years. Within the last 20,000 years, much of northern North America, northern Europe, and northern Asia was covered by continental glaciers. This last phase of the Pleistocene, called the late Wisconsinan Glaciation, began more than 125,000 years ago. This glacial event is just one of many natural climate variations that Earth has experienced. With the significant increase in human population over the last few centuries and with the coming of the Industrial Revolution, human activities may also have strongly influenced climate, a trend that continues today. Unlike the northeastern and northcentral United States, where continental glaciers occurred only during the last two million years, parts of Alaska have possessed the right combination of climate and geology to support almost continuous glaciation for at least the last five million to 12 million years. Glaciologists think an even older, extensive glacial history existed in Alaska, but the record has been obscured by intensive, more recent glaciations.

Climate models, for example the General Circulation Models (GCMs) developed to predict the impact of increased greenhouse gases such as carbon dioxide (CO_2) and methane (CH_3) on Earth's climate, foresee a warming trend for northwestern North America, including Alaska. For other areas such as the eastern Canadian Arctic and western Greenland, the models predict cooling trends or no change.

Based on temperature dynamics, scientists recognize two types of glaciers: polar and temperate. A climate change, be it natural or human induced, involving a warming of several degrees can significantly change the thermal regime of temperate glaciers, resulting in increased melting. Such a warming was noted in Alaska for much of the twentieth century.

Today, glaciers cover about 29,000 square miles, about five percent of Alaska. Less than 20,000 years ago, during the last glacial maximum, nearly 300,000 square miles, about 50 percent of Alaska, and much of the continental shelf surrounding the state lay shrouded by large glaciers, icefields, ice caps, and subcontinental ice sheets. A huge ice sheet covered the northern Gulf of Alaska, burying Middleton Island, and may have had a floating, iceberg-calving terminus similar to the present-day Ross Ice Shelf in Antarctica. Much of the ground in the glacier-free parts of Alaska was perennially frozen, a condition that still persists in many northern areas and is known as permafrost.

Seventy-five miles west of Anchorage, Mount Gerdine overlooks Triumvirate Glacier, so named because of its three large tributaries in the Tordrillo Mountains. This photo shows several aspects of a glacier, including lateral and medial moraines, nunataks, trimlines, icefalls, and crevasses. (Fred Hirschmann)

About 15,000 years ago Alaska's massive ice cover began to melt. Many of the strongholds of glaciation that exist today have endured since the time of the gigantic Pleistocene ice cover. The late Wisconsinan Glaciation ended between 10,500 and 12,500 years ago. By about 10,000 years ago, glacier coverage was not too much different from today. Since then, a series of climatic oscillations has occurred, involving both warming and cooling.

During the Holocene epoch, or modern times, parts of Alaska have experienced several separate episodes of cooling and glacial readvance. At different times, these have affected glaciers from the Alexander Archipelago, in Southeast, to the Brooks Range. About 5,300 to 6,600 years ago, the climate was even warmer than present, perhaps by as much as three to five degrees Fahrenheit. This interval, during which glaciers retreated throughout Alaska, is referred to as the Altithermal. The last major cooling event, called the Little Ice Age, began about 650 years ago. A global natural climate cooling event, it caused most of the temperate glaciers of North and South America, Africa, Europe, Asia, and on the islands of New Zealand and New Guinea to expand to their largest size since the end of the Pleistocene. A dramatic expansion of the area, thickness, and volume of many Alaska glaciers took place. Retreat began from 200 to 300 years ago. Today, many Alaska glaciers are only now retreating from their Little Ice Age maximum positions. Other, similar periods of glacial advance occurred 7,000 to 8,200 years ago, 4,900 to 5,300 years ago and 2,400 to 3,300 years ago. The collective series of glacial advances is referred to as the Neoglaciation. During non-glacial times, climate was significantly warmer.

The future of glaciers and glaciation stimulates much speculation. Today most Alaska glaciers are thinning and retreating in response to a complex regional climate change. The media uses the term "global warming" to describe this change. However, what is happening is far more complex than a simple increase in global temperature.

What does the future hold? Only a minor increase in precipitation and decrease in temperature is necessary to reverse current changes and cause significant buildups in the snow packs of existing Alaska glaciers. This could easily produce the beginnings of significant glacier expansion and advance. Continued warming and decreased precipitation, on the other hand, could prolong the shrinking trend of most of the state's glaciers and ultimately lead to the disappearance of numbers of them.

LEFT: *At Sitkagi Bluffs, the stagnant-ice terminus of Malaspina Glacier, a mature spruce forest grows on a veneer of glacial sediment covering the ice. The Pacific Ocean lies beyond. (Bruce Molnia)*

FACING PAGE: *An old trimline, visible along a valley wall in the Chigmit Mountains, Lake Clark National Park, suggests the bulldozing power behind glacier ice. This glacier feeds the Tlikakila River, designated a national wild and scenic river, which flows into Lake Clark. (Chlaus Lotscher)*

Glaciers and Alaska's Ice-Age Fossils

By Paul Matheus

EDITOR'S NOTE: *Paul Matheus, director of the Alaska Quaternary Center, wrote about ice-age mammals in* Prehistoric Alaska, ALASKA GEOGRAPHIC® *Vol. 21, No. 4.*

People commonly think animals and plants get preserved directly in glacier ice. As recently as 50 years ago, the popular press portrayed ice-age beasts trapped by advancing Pleistocene glaciers, including mammoths frozen in massive ice with a last meal of buttercups hanging out of their mouth. As provocative as these images sound, they are long-standing myths. Except in rare cases, such as Otzi, the Italian Ice Man, found in 1991, it turns out that glaciers themselves seldom function as direct agents of fossil preservation. However, climatic conditions during the ice ages and associated periglacial environments are indeed responsible for preserving many fossils of Pleistocene animals and plants. Moreover, climatic conditions of glacial periods in the Pleistocene led to development of plant and animal communities in Alaska far different from the familiar tundra and taiga communities of today.

Periglacial environments, those adjacent to and shaped by glaciers, are good for preserving fossils for two main reasons. First, and most obvious, is the fact that cold conditions slow or even halt decomposition of organic materials. This gives the remains of animals and plants a chance to be buried and preserved before totally decomposing. Second, and perhaps more important, glaciers grind rock into fine silt that is picked up and deposited by periglacial winds. This wind-blown silt, called loess, is a good agent for burying and preserving bones because it forms a high pH (non-acidic) packaging material, especially in places like central and northern Alaska. The pH of loess depends on the parent rock from which it is ground, but much of the parent rock in Alaska, especially from the Alaska and Brooks Ranges, is calcareous, and calcareous sediments neutralize naturally occurring acids that would otherwise decompose bones. For fossils to be preserved, they need to be buried fairly soon after death by significant amounts of sediment, and in conditions that prevent decomposition. Periglacial environments do this well.

Bones of late-Pleistocene mammals (10,000 to 50,000 years old) tend to be so well preserved in Alaska and the Yukon that most still contain their original organic material and structure. Unlike dinosaur bones, most Pleistocene bones have not been significantly altered by minerals permeating into them from the environment. Some of these frozen bones still contain bone marrow in their cavities, and fragments of original DNA. By studying the DNA and chemical components of Pleistocene mammal bones, researchers have been able to gather information about the genetic structure and diets of ancient mammal populations in

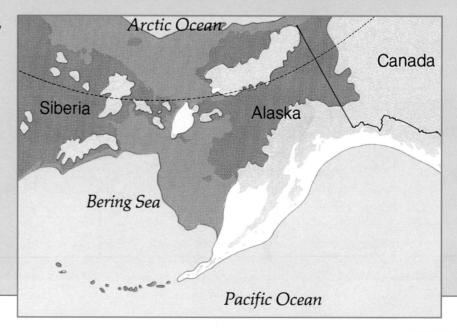

This map shows the extent of glaciation during the Pleistocene, when the Bering Land Bridge connected Alaska and Siberia. (ALASKA GEOGRAPHIC® map, modified from USGS)

Granite glacial erratics *bigger than a house sit where a former glacier dropped them near Big River on the north side of the Alaska Range. These huge blocks were transported down the glacier as if on a conveyor belt and stranded on the valley floor miles from their origin when the ice melted. Fossils may also be ice-rafted away from their original position. (Marydith Beeman)*

Alaska. This information helps scientists understand when and how animals migrated into Alaska through the late Pleistocene and why many went extinct between about 10,000 to 12,000 years ago.

During the Pleistocene, Alaska experienced alternating episodes of glacial and interglacial periods. *Glacials* tend to last on the order of 100,000 years and *interglacials* 15,000 to 20,000 years. Frequently, there are periods of moderate warming within glacials, called *interstadials*. During interglacial periods, such as today, boreal forest develops throughout much of Alaska. During the last interglacial period, 115,000 to 130,000 years ago, boreal forest even grew on most of the North Slope (the last interglacial was approximately three to six degrees Fahrenheit warmer than today). During interstadial periods, an open forest woodland with boreal tree species develops in much of Alaska south of the Brooks Range. The most recent interstadial period occurred from about 30,000 to 60,000 years ago. Glacial periods in Alaska are cold and arid, leading to the formation of mostly treeless ecosystems dominated by sparse grasslands and dry tundra. A rough analog would be certain cold, arid steppes of Siberia, Mongolia, and northern China. These cold, dry ecosystems occurred during much of the time before and after the last interglacial period when glaciers extended over large portions of the Northern Hemisphere. The most recent glaciation, known as the Wisconsinan Glaciation, occurred from about 12,000 to 125,000 years ago, but was interrupted by the above-mentioned interstadial period.

Alaska's geography — and consequently its biogeography — have been significantly affected by the Pleistocene cycle of glaciation. During glacial periods, much of Earth's water becomes tied up as glacier ice, and global sea level drops as much as 300 feet. Alaska's shallow oceanic shelves in the Bering and Chukchi Seas become dry and form a broad, lowland plain. This Bering Land Bridge connected North America to Asia via Alaska during glacial periods, and has served as the only terrestrial route for animal and plant migrations between the Old World and New World since the Eocene period about 45 million years ago, when North America and Europe were connected via Greenland. Of course, high sea levels during interglacial periods inundated the land bridge.

Since the vegetation of Alaska varied so dramatically between glacial and interglacial periods of the Pleistocene, so too did its large mammal fauna. Mammal fossils from interglacial periods are rare because forest conditions are not conducive to bone preservation. Still, limited evidence suggests that during interglacials Alaska was inhabited by forest-dependent mammals such as mastodons, tree squirrels, martens, and black bears. Scientists suspect that areas of tundra and broken forests were home to bison, caribou, mammoth, horse, and a few different species of muskox. During glacial periods, a distinct grazing (grass-eating) fauna predominated. During the Wisconsinan Glaciation (researchers know little about older glaciations), mammoth, bison, and horse accounted for more than 80 percent of the large mammal biomass in Alaska. Caribou and muskox also are common fossils from glacial times. Renowned ice-age carnivores included lions (essentially the same lion as in Africa today), wolves, short-faced bears (giant, lean bears that were strictly flesh-eaters), dirk-toothed cats (a type of saber-tooth), wolverines, and brown bears.

Paleontologists have discovered a few fossils of rare mammals from Pleistocene Alaska that are of uncertain age and association. Among these is the giant ground sloth, a huge beast that still has biologists perplexed as to how it made a living. A couple of bones have been found of a coyote-sized, highly social canid called the dhole. Dholes are still extant in India and central Asia. The stag moose, a relative of the modern moose, seems to have inhabited Alaska perhaps as late as the Wisconsinan Glaciation. Modern moose entered Alaska from Asia around 12,000 years ago.

Although somewhat counter-intuitive, today's large mammal community is comparatively less diverse than it was during colder, drier glacial times. It even appears to be less diverse than during the previous interglacial. The reason is

In 1906 (above), a glacier filled this valley near Mount McKinley. Eighty-nine years later, however, the ice had melted. Since cold conditions contribute to fossil preservation, the current warming trend may expose ancient plant and animal fossils.

By 1995 (right), the glacier had disappeared, leaving only its imprint against the valley walls. Alternating periods of cooling and warming have been occurring in Alaska for thousands of years, creating and melting glaciers accordingly. (Above, courtesy Frederick A. Cook Society; facing page, Brian Okonek)

because today's Alaska ecosystem — boreal forest and tundra — offers relatively little in the way of good-quality, digestible forage for large herbivores. Caaribou and moose manage to survive today through unique adaptations and foraging ecologies — they are the exceptions. Grasslands, even cold, dry grasslands, provide inherently more digestible forage for large herbivores. So, even though the climate in Alaska was harsher during the ice ages, conditions were actually better for most large mammals. The interstadial period from 30,000 to 60,000 years ago may have been the best of both worlds. Apparently, open woodlands and grasslands of the interstadial supported nearly all of the species listed above.

If glaciers are not direct agents of fossil preservation, then why do people occasionally find frozen mummies of ice-age mammals in Arctic and Subarctic regions of Alaska, Siberia, and the Yukon?

Frozen mummies are not the result of animals being trapped in glaciers, but form by a process of desiccation when animals are buried in sediments that subsequently become frozen as part of the permafrost zone. Moisture is drawn out of the animal's body via natural osmotic movement (segregation) of water molecules from the animal to drier sediments that surround it. The process can take as little as a few decades to hundreds or thousands of years, depending on the surrounding moisture levels. The animals recovered from snow patches in Yukon Territory (see page 42) have been preserved by this process, except that the surrounding "sediments" were snow rather than dirt. It is not uncommon for mummies to be found with ice surrounding them, thus leading to the misconception that they were flash-frozen by glacier ice. In fact, the ice surrounding mummies forms as soil segregation ice, with some of the water no doubt coming from the mummies themselves. Mammoth, horse, bison, ground squirrel, helmeted muskox (extinct), caribou, moose, voles, pika, snowshoe hare, and lynx mummies have been recovered.

Mummified remains of a bison, mammoth, and ground squirrel are on display at the University of Alaska Museum. Parts of a horse mummy can be viewed at the Beringia Interpretive Centre in Whitehorse, Yukon Territory. The Beringia Centre also has many fantastic skeletal reconstructions of ice-age mammals and an amazing fleshed-out scimitar cat that paleontologists and artists reconstructed by modifying a taxidermied lion.

Although much of Pleistocene Alaska was glaciated, those areas not covered by ice offered rich environments for Alaska's menagerie of prehistoric mammals.

Glacier Mass Balance: Accumulation versus Ablation

Much like a bank account, a glacier registers deposits and withdrawals. Deposits accumulate by new snow falling on its surface, avalanches releasing ice and snow from its valley walls, and snow blowing in from adjacent basins. Withdrawals are made by melting, evaporation, sublimation, and in the case of glaciers that end in water, by calving. When more snow and ice build up in the accumulation area than is lost in the ablation area, scientists refer to a glacier as being "healthy," with a positive budget. Such glaciers are generally thickening and frequently advancing. Glaciers dominated by a negative ice budget usually are thinning and retreating. This category of glaciers is euphemistically referred to as being "sick."

The surface of a glacier can be divided into two zones: the zone of accumulation, where more snow accumulates most years than is lost to melting, evaporation, or sublimation; and the zone of ablation, where more snow and ice is lost to these processes than accumulates. The line where these two zones meet, where the annual loss is equal to the annual gain, is the equilibrium line, which is often close to the firn line, the lower limit of year-round snow cover.

A glacier's health has a significant influence on its length and thickness. Healthy glaciers thicken, resulting in more ice being available to be transferred down-glacier. A delicate balance exists between the rate that ice moves down-glacier and the rate that melting or calving is taking place. If ice loss at the terminus exceeds replacement, the distance between the head of the glacier and its terminus decreases, and the glacier is said to be retreating. If the distance between the head of the glacier and its terminus increases, the glacier is advancing. Actually, until it becomes stagnant, the ice in a glacier always moves down-slope, even when the terminus of a glacier is rapidly retreating. If the quantity of ice lost to melting and calving exceeds the amount of ice coming from above, then the position of the glacier's terminus will retreat. When the supply is greater than the loss, then the terminus will advance. When supply is equal to loss, then the position of the terminus remains stationary.

As Meares Glacier advances, it uproots mature spruce trees and bulldozes the ground in its path. Meares flows from the Chugach Mountains to Unakwik Inlet about 40 miles west of Valdez. (Bruce Molnia)

Monitoring and Measuring Alaska Glaciers

Glacier monitoring methodically documents changes in a glacier system. These include many types of changes. For instance, glaciologists may be interested in changes in the mass balance of a glacier and may make seasonal measurements to determine snow accumulation and ice melt. They also document changes in the position of the glacier's terminus or of other features on a glacier's surface, such as the firn line or an avalanche deposit. Glacial ecologists may want to monitor the rate and sequence of plant succession on sediment covering a glacier; a glacial geologist may strive to decipher the recent history of a glacier system and document the chronology of change. All may be interested in changes in area and thickness. Providing answers to these and similar questions involves a combination of hands-on field sampling and use of a broad spectrum of remote-sensing devices to provide a different perspective and scale of information than can be sampled on a glacier surface. In each case the precise location of the features being studied is as critical as accurate field measurements. Until recently, pinpointing a feature's position was the weak link in Alaska glacier monitoring.

In the last few years, growth of the Global Positioning System (GPS) has revolutionized nearly every type of glacier measurement. The GPS, developed by the U.S. Department of Defense, is a worldwide, satellite-based, radio-navigation system that provides positional accuracy of less than an inch. It became operational in 1995. Originally created for military use, today it is available to everyone. Sophisticated GPS receivers are small enough to be hand-carried and provide real time measurements of latitude, longitude, elevation, time, distance from a reference point, and bearing. They can outline the shape and position of a glacier margin or precisely locate any feature on the surface of a glacier in three dimensions. Repeat measurements at the same location help determine subtle changes in elevation produced by isostatic changes.

Prior to GPS, determining the rate of ice flow at the surface of a glacier demanded a variety of surveying techniques that involved either measurement of distances between a point on the glacier's surface and a fixed survey point, or measurement of angles between fixed survey points and stakes or flags placed on the ice surface. For positional accuracy, these measurements required a network of regional benchmarks, usually significant distances from investigation areas. Measurements were made over a period of time to accurately determine movement rate. Instruments used in these determinations, aladades, theodolites, and photo-theodolites, technologies from the late nineteenth and early twentieth centuries, were weather-dependant. With GPS, a sensor can be left on the glacier's surface that continuously measures and records position and can even transmit information to a remote site for collection and computation.

Another late-twentieth-century technology for measuring glacier movement is

Rachel and Tasha Syverson traverse an ashy basin beneath a glacier at the foot of Mount Mageik, a volcano flanking the Valley of Ten Thousand Smokes in the Aleutian Range. A catastrophic eruption in 1912 created this barren landscape. Most glaciers here are unnamed. (Greg Syverson)

Doppler laser. With a Doppler laser system, one needs only 50 seconds to accurately determine the velocity of a glacier that moves at a rate of as little as five feet per year. Rates of movement as small as .000001 inches per second can be measured at distances of more than a mile. The only drawback to this type of instantaneous measurement is that the period of observation is so short that daily variations in velocity cannot be considered.

Since 1993, scientists from the University of Alaska Fairbanks have used GPS and a laser-ranging system mounted in an aircraft

Bearberry bushes turn red each fall above Exit Glacier, accessible by road a few miles from Seward. A footpath with signs marking stages of ice movement, leads over old moraines to the terminus. (Jon R. Nickles)

to profile the surface of about 70 glaciers throughout Alaska. Computer software processes the GPS data to provide precision location information for the aircraft. The laser profiles of a glacier's surface are then located with respect to the aircraft's position with an accuracy of about one foot. Researchers compare their profiles with USGS topographic maps, made in the 1950s, to determine changes in the glacier's surface elevation. Their results show that some glaciers have thinned by more than 150 feet.

Some questions involve determining how the location of a glacier's terminus has changed with time. For example, how did the position of glacier termini in a particular region change during the twentieth century? These types of questions are answered by comparisons of ground, aerial, and space-based photographs or other remotely sensed imagery. The photographic record of Alaska glaciers began in the 1880s. As early as 1926, aerial photography of southern Alaska was being collected. So for certain areas an accurate baseline of photographic-derived information can be constructed for much of the early twentieth century. Photographic methods can have a resolution, i.e. distinguish objects, of about one foot. Airborne photographic sensors are still an important source of twenty-first-century information about glaciers. Since the early 1970s, aerial photography has been supplemented by airborne and space-based sensors, such as Landsat, synthetic-aperture radar (SAR), and digital space photography. Combinations of these provide information, ranging from broad regional coverage to large-scale detailed information about small features.

One of the oldest and most reliable digital satellite systems is Landsat, currently operated by the U.S. government. This program, which began in 1972, has polar-orbiting satellites that regularly circle over Alaska and transmit digital information back to Earth. Although the maximum resolution is only about 50 feet, large changes such as surges and glacier lake floods can easily be measured. Several commercial satellite systems provide both digital imagery and photography from space. One of the newest has a resolution of about three feet. The greatest limitation of these commercially operated systems is the cost to purchase data.

Commercially operated digital radar systems also receive imagery of Alaska glaciers. Radar employs microwave energy, which permits measurement and depiction of Earth's surface features regardless of weather conditions. Radar systems used today are of the SAR type, a system that allows much higher resolution than earlier real-aperture types. For instance, X-band SAR has a wavelength of one inch and a resolution of three feet. SAR is one of the sophisticated tools enlisted in the detailed investigation of many Alaska glaciers. One SAR technique known as interferometry can detect subtle movement of individual features on a glacier's surface from hundreds of miles in space.

A ground-based, ice-penetrating radar technique, also called radio-echo sounding, permits measurements of glacier thickness and shape. This system uses a wavelength that is several feet long and requires that an antenna be placed on the glacier's surface.

Red Glacier, one of several on Mount Iliamna, an active volcano on the west side of Cook Inlet, supports plantlife. Plant succession on recently deglaciated terrain depends on factors such as latitude, altitude, and temperature. (Bruce Molnia)

Out of the Northern Ice

By Greg Hare

Ancient hunting artifacts such as this spear used to hunt caribou are appearing from melting icefields in Yukon Territory near Alaska's border. Canada's First Nations people are cooperating with scientists to gather and care for these treasures. (Courtesy Government of Yukon Territory)

EDITOR'S NOTE: *Greg Hare of Whitehorse, an archaeologist with the Heritage Branch of the Yukon territorial government, wrote this article about important recent finds in northern glacial archaeology. Although an ice patch treasure trove has yet to be discovered in Alaska, environmental conditions like those in neighboring Yukon Territory exist in the eastern part of the state and scientists think it will only take some looking to find similar prehistoric records.*

In August 1997, a wildlife biologist hiking through the mountains of southwest Yukon Territory encountered a portal to the past.

While hunting sheep, Gerry Kuzyk and his wife, Kristin Benedek, noticed a large alpine ice patch covered in dark, and highly pungent, pellets of caribou dung. These pellets were nearly a foot thick in places, but Gerry knew that no caribou had been seen in this area for nearly 60 years.

Gerry later returned to the site with another caribou scientist to confirm the identification. But they were perplexed. How could there be so much caribou dung, where there are no caribou? During that visit, they picked up a few frozen pellets and a small brown stick with a piece of sinew attached.

In an effort to solve the mystery, a caribou pellet and small sample of the stick were sent for radiocarbon dating. Those two dates revealed that the dung was many thousands of years old, and the stick was actually the remains of a 4,500-year-old hunting dart.

Since that initial discovery, dozens of ancient alpine ice patches have been found in southwest Yukon, the oldest more than 8,000 years old. But they now appear to be melting rapidly. As they melt, they are revealing not only the accumulated dung of the caribou that used the ice patches for millennia, but also an unprecedented assemblage of ancient hunting artifacts, faunal remains, biological specimens, and other sources of paleoenvironmental information.

Over the past three years, a variety of multidisciplinary and cross-cultural research efforts have begun, involving government, university, and First Nations researchers and managers. Their shared objective is to discover the clues to the past still frozen in the Yukon ice.

This is the traditional homeland of four different Yukon First Nations: Champagne and Aishihik, Carcross-Tagish, Kwanlin Dun, and Kluane. From the outset the First Nations have been involved in the research and are today co-managers of the project along with the Yukon Government Department of Renewable Resources and the Heritage Branch.

Most of the ice patches lie in the rugged Coast Mountains and Ruby Range of southwest Yukon. Many are visible from the Alaska Highway west of Whitehorse in August. At a distance they look like any other snow-capped mountains, but upon closer inspection, many are seen to contain the characteristic layers of ancient caribou dung.

Hunters and biologists have long observed that in July and August woodland caribou congregate on alpine ice patches to seek relief from insects and summer heat. However, before 1997 no one realized that these ice patches had persisted for thousands of years. In times gone by, these reliable gatherings of caribou must have served as the local supermarket for Native hunters — with a built-in deep freeze.

Through the millennia, ancient hunters armed with spears, darts, or bows and arrows hid behind boulders or constructed their own stone blinds, casting their projectiles at the lethargic caribou. When they missed their target, their darts often disappeared into the soft summer snow. Today, these ice patches are melting, their permanence undermined by the late twentieth century's mild winters and hot summers. Hunting weapons that have not seen the light of day for hundreds, even thousands of years, lie exposed on the glistening surface of melting ice surrounded by thousands of decaying caribou pellets.

First Nations researchers and archaeologists have recovered more than 100 examples of ancient hunting artifacts, ranging from six-foot-long throwing darts to recent arrows, complete with barbed antler tips, feathers, sinew, and ocher markings. Radiocarbon dates run on a number of artifacts range from more than 7,000 years old to the end of the nineteenth century.

But the artifacts are only a small part of the incredible story contained in the ice. The caribou

Caribou herds have roamed the wilds of Alaska and Canada for centuries, oblivious to international borders. Melting ice patches have recently revealed layers of old caribou dung in areas of Yukon Territory where no herds currently exist. Similar discoveries might also be made in Alaska in the future. (James L. Davis)

dung itself has proven to be a tremendous source of information. From the pellets, researchers are working to extract information about caribou diet and health and genetic changes through time. Pollen and plant remains found in the pellets make it possible to reconstruct past environments and climates. Even parasites, some thought to have been recent arrivals in the North, have been shown to be present in caribou 5,000 years ago.

Also preserved in the ice have been the carcasses of more than 100 small mammals and birds, some which may be thousands of years old. Many of these small mammals are sensitive indicators of environmental change. Already several of the recovered animals challenge the known geographical range of their species. Other preserved mammals may cast light on the development of viruses such as hanta. This faunal collection is currently archived at the University of Alaska Museum Frozen Tissue Collection in Fairbanks, and the Alaska Quaternary Center is conducting stable isotope analysis on faunal material from the ice patches. This analysis helps to reveal differences in the ecology, physiology, and diets of various species and to reconstruct specific predator-prey relationships.

For the First Nations, the significance of this project goes beyond the science. It provides not only opportunities for student involvement and education, but integrates the knowledge of traditional hunting strategies and land use into the scientific interpretation of the ice patch phenomenon. This past summer saw Champagne and Aishihik First Nations organize for the first time a week-long science camp aimed at introducing First Nations youth to aspects of the Ice Patch Project and linking science and archaeology with the community.

Research into melting alpine ice patches is still in the developmental stages. Each year the scope and boundaries of the project expand, and each spring finds researchers eagerly awaiting the warm winds of summer to see what the ice will yield.

Lowell Glacier weaves through the St. Elias Mountains in Kluane National Park, a spectacular backdrop for wild mountain goats. Several glaciers in this range originate in Canada but flow southwest into Alaska. (Michael R. Speaks)

Effects of Glaciers on Sea Level and Earth's Crust

It's a matter of simple addition and subtraction. For glaciers to expand in size, precipitation, fueled by water evaporating from Earth's oceans, must increase. As glaciers expand, sea level decreases. Similarly, as glaciers expand and thicken, more weight presses on the crust below. In response, the crust deforms. These processes are called **eustacy** and **isostacy**. Both cause base level changes in glacial environments.

Eustacy describes the worldwide sea level regime and its fluctuations caused by changes in the quantity of seawater available. During the Pleistocene, growing continental ice sheets tied up so much fresh water that worldwide sea level decreased by about 300 feet. In Alaska, eustatic sea level lowering left many present-day fiords isolated from the Pacific Ocean. Many glacial erosion features, such as the Yakutat Sea Valley and the Bering Trough, both deep channels cut into the exposed continental shelf of the Gulf of Alaska, were formed by glaciers that advanced onto the exposed continental shelf. As sea level rose, these features were drowned.

A further growth of glaciers would cause another episode of eustatic sea level lowering. As a result of late-Pleistocene eustatic sea level lowering, Earth's land area increased by eight percent as the continental shelves of the entire world, not only in glacier-covered areas but everywhere, were exposed. During the last glacial maximum, 37.1 percent of the planet was exposed land compared to only 29.1 percent today. Between 12,000 and 15,000 years ago, when the major continental ice sheets began to melt, sea level started to rise. Equilibrium at near the present level was reached only about 6,000 years ago.

Today's changing climate worries many people. What would the impact be, particularly on areas of extremely low relief such as many Pacific Islands, Bangladesh, and even Florida, if temperate glaciers, especially Alaska's, melted? Sea level would rise a little less than a foot if just Alaska's glaciers melted. On the other hand, melting of the Antarctic and Greenland ice sheets, where the majority of glacier ice resides, would result in an additional eustatic sea level rise of about 265 feet, flooding every coastal city on every continent. Sea level would rise about 243 feet if only Antarctica's ice melted and about 22 feet from just Greenland's

Where meltwater from Malaspina Glacier enters the Gulf of Alaska, the flow of glacial and non-glacial water becomes apparent. Silty gray patches eventually mix with clearer seawater. (Loren Taft)

melted ice cover. Melting of just the West Antarctic Ice Sheet, the Antarctic area frequently described as being vulnerable to changing climate, would cause a sea level rise of about 26.5 feet.

Isostacy describes changes within the planet's surface where material in the crust and mantle is displaced in response to the increase (isostatic depression) or decrease

LEFT: *Ancient trees, having been snapped like twigs and carried along in the ice, began emerging from under Bering Glacier in summer 2000. (Bruce Molnia)*

ABOVE: *Grasses, flowers, and shrubs are the first plants to recolonize a deglaciated area. These were found near Baird Glacier, in the Tongass National Forest. (Pieter Folkens)*

(isostatic rebound) in mass at any point on Earth's surface. As a result of an increase in mass of material caused by the presence of hundreds to thousands of feet of glacier ice overlying the ground surface, many areas of Alaska were depressed. East of Alaska, Hudson Bay was the center of the large ice sheet that covered eastern North America. Earth's crust in the area of the bay was depressed by more than 600 feet. Subsequent melting of the ice mass resulted in the slow rising of the ground surface through isostatic readjustment. This rebound continues. In Alaska, parts of the crust may have been depressed by as much as 200 feet. The Bartlett Cove area of Glacier Bay has risen an average of about one inch per year since it was deglaciated about 200 years ago. This rate of isostatic readjustment is also valid for many other Alaska areas.

A recent study showed that along the Gulf of Alaska coastline near Bering Glacier, 0.04 to 0.5 inches of isostatic rebound occurs annually, driven by the retreat of Bering Glacier. However, the 1993-1995 surge of Bering Glacier caused the coastal plain in front of it to be depressed by nearly an inch. As the glacier has retreated from its surge maximum position, the coast has begun to rise. Similarly, the Gulf of Alaska coastline adjacent to Icy Bay may have been uplifted about three feet during the twentieth century.

Exploration of Alaska Glaciers

Several volumes could be written on exploration of Alaska and development of knowledge about the state's glaciers. A few highlights are presented here to emphasize important milestones. The western world knew little about glaciers until the middle of the nineteenth century, hence it is not surprising that many of the classic exploration voyages of the eighteenth century failed to recognize their presence in southern Alaska. However, Alaska's indigenous peoples knew about the existence of glaciers. Groups such as the Tlingit lived near glaciers and developed a culture that recognized the influence glaciers played on their daily activities. Traditional knowledge records instances of Native villages being destroyed by advancing glaciers.

Nearly all European naval explorers of the late eighteenth and early nineteenth centuries came from countries without modern glaciers. Capt. Don Alessandro Malaspina, Capt. George Vancouver, Lt. Peter Puget, and even Capt. James Cook thought that the walls of ice and snow they found were either sea ice or part of the northern polar barrier and not glaciers. Their journals refer to areas of perpetual frost, floating compact bodies of ice, and floating masses of snow, but none describe the glaciers they saw. In 1778, Cook and Vancouver sailed into Cross Sound past the gigantic glacier that filled the mouth of Glacier Bay and failed to identify it as a glacier. The first description of glacier calving in Alaska was made during Vancouver's 1794 exploration of College Fiord in Prince William Sound. He describes the frightening, thunderous roars the crew frequently heard and the blocks of frozen snow falling off the snow cliff faces.

> What is that roar or explosion that salutes our ears before our anchor has found bottom? It is the downpour of an enormous mass of ice from the glacier's front, making it for the moment as active as Niagara.
> (1899, John Burroughs, *Alaska: The Harriman Expedition*)

Of the eighteenth-century explorers only Jean Francois de Galaup de La Perouse recognized the significance of the glaciers he observed in 1786. La Perouse's map of Lituya Bay, published in 1797, accurately depicts the location of five glaciers in the upper ends of the bay. The accuracy of La Perouse's map provides scientists with a base from which to

Nestled in the U-shaped South Fork valley of Eagle River north of Anchorage, two lakes demonstrate different water sources. Separated by an old medial moraine, Symphony Lake, left, is fed by a clearwater stream, while the milky color of Eagle Lake belies its glacial origin. Only five miles from the nearest trailhead, this is a popular destination for hikers. (Jon R. Nickles)

compare changes in the terminus positions of Lituya Bay's glaciers over more than two centuries.

Little was written about Alaska's glaciers during the first two-thirds of the nineteenth century. However, between 1848 and 1850 a series of charts and maps appeared that was prepared by Kozima Trentier and Mikhail Kadin for Admiral Mikhail Dmitrievich Teben'kov, director of the Russian American Co. and governor of Russian America. These maps, published as the *Atlas of the Northwest Coast of America*, showed the location of many coastal features including glaciers. The information presented was the result of more than 100 years of observations by Russian explorers beginning with Vitus Bering in 1741.

In 1863, the Russian Naval Squadron invited Professor William Blake, an American, to accompany the crew on an exploration of the Stikine River Valley. Blake's investigation resulted in the 1867 publication in the *American Journal of Science* of "Glaciers of Alaska, Russian America," the first scientific summary of Alaska glaciers. Between 1867 and World War I, much of the knowledge resulted from U.S. government surveys, reports of climbing expeditions, and the privately funded observations of a few naturalists and geologists.

Rising seemingly from nowhere, these glacial erratics are leftovers from past glaciations near park headquarters in Denali National Park. Presently, cirque glaciers, valley glaciers, piedmont glaciers, and icefields exist in Alaska. Ice caps, ice sheets, and ice shelves, all of which are larger, do not occur in the state today, though they may have in the geologic past. (Curvin Metzler)

> The snow on the [Ruth] glacier was hard and offered a splendid surface for a rapid march but the advantage of its hardness was offset by the treacherous manner in which it bridged dangerous crevasses. As we advanced these snow bridges increased and we held to our horsehair rope with more interest.
>
> (1906, Frederick A. Cook, *To the Top of the Continent*)

John Muir, a naturalist and travel writer who later founded the American environmental movement, visited Alaska for the first time in 1879. Accompanied by Samuel Hall Young, a Presbyterian missionary from Sitka, Muir explored Glacier Bay and Muir Glacier. In later visits in 1881 and 1899, he noted changes in the bay and explored numerous glaciers in Yakutat Bay, Prince William Sound, and Cook Inlet. Muir's writings were popular and easily available, and stimulated others to explore Glacier Bay, including Henry Fielding Reid of Case School of Applied Science in Cleveland, who mapped Glacier Bay in detail in 1890 and 1892.

In 1890 and 1891, USGS geologist Israel Cook Russell led an expedition, jointly funded by USGS and the National Geographic Society (NGS), to explore the geology of the Yakutat Bay region. Its primary objective was to climb 18,008-foot Mount St. Elias. Although severe weather foiled the ascent, Russell explored much of Malaspina Glacier, Yakutat Bay, and Russell Fiord. Russell's expedition produced the earliest photographs of Yakutat Bay, Malaspina Glacier, and the Mount St. Elias region.

Following Russell's expedition, NGS became a focal point for both organizing and funding studies of Alaska glaciers and for reporting their results in its magazine. From the early 1890s through the 1920s, they presented the results of Frederick Schwatka's and C.W. Hayes's explorations of the Copper River and the area north of the St. Elias Mountains; Reid's descriptions of Glacier Bay and Muir Glacier; Henry Gannett's description of Alaska glaciers, including those in Prince William Sound; E.R. Skidmore's descriptions of the glaciers of the Stikine River area; C.L. Andrews's and Fremont Morse's studies of Muir Glacier during the first decade of the twentieth century; W.C. Mendenhall's exploration of Wrangell Mountains glaciers; studies of Mount McKinley region glaciers by W.A. Dickey, Alfred Hulse Brooks, and Robert Muldrow; W.H. Osgood's descriptions of Lake Clark region glaciers; and Ferdinand Westdahl's photographs of glaciers in the Aleutian Islands.

In 1897, Prince Luigi Amedeo Di Savoia, Duke of the Abruzzi, successfully led an expedition to Mount St. Elias's summit, naming many of the glacial features that were seen during the ascent. What made his expedition unique was that his support team carried a large panorama camera to the top of the mountain, where they produced numerous panoramas of Malaspina Glacier, upper Bering Glacier, and the central St. Elias Mountains.

The Harriman Alaska Expedition of 1899, privately funded by Edward Henry Harriman as a vacation for his family and as a multidisciplinary scientific study, was the most prolific scientific exploration of Alaska of its time. A dozen volumes of scientific observations were published as a result. Harriman chartered the steamer *George W. Elder* and invited about two dozen distinguished scientists and naturalists to sail to Alaska. Among them were John Muir, William H. Dall of the Smithsonian Institution, Benjamin K. Emerson of Amherst College, Charles Palache of Harvard University, and Grove Karl Gilbert of USGS. Edward S.

An ice chunk from the tidewater terminus of Hubbard Glacier, at the head of Disenchantment Bay north of Yakutat, crashes into the sea, creating a splash wave forceful enough to overturn boats that venture too close. (R.E. Johnson)

Sunrise over Matanuska Glacier turns surrounding peaks and those of the distant Chugach Mountains pink. Animal tracks dot the silty mud in the lower left corner of this image. (Fred Hirschmann)

> The box was set on the bow for a moment, just after landing, and then the glacier calved. It was big, and as I turned to look the swell from the preceding explosion hit the boat and all our food went over the side. Surrounded by floating grapefruits I shouted and cursed at the Grand Pacific, but it kept right on calving. It paid no attention to me. This country doesn't pay attention to anybody.
>
> (1965, Dave Bohn, from the anthology *A Republic of Rivers: Three Centuries of Nature Writing from Alaska and the Yukon*)

Curtis, famous later for his photographs of American Indians, was the expedition's official photographer. Scientific disciplines covered by the expedition included ethnology, zoology, botany, geography, and geology. The expedition sailed from Seattle on May 31, 1899, traveled more than 9,000 miles, and made about 50 scientific stops. Among the Alaska locations visited were Annette Island, Wrangell, Juneau, Glacier Bay, Sitka, La Perouse Glacier, Yakutat Bay, Prince William Sound, Cook Inlet, Homer, Kodiak, Dutch Harbor, Bogoslof Island, St. Paul and St. Matthew Islands, and Cape Fox. The expedition returned to Seattle on July 30, 1899. Detailed observations in Glacier Bay and Yakutat Bay expanded previous knowledge. In Prince William Sound, the expedition explored the glaciers of College Fiord and discovered a completely unknown inlet, dubbing it Harriman Fiord. Two volumes were published that summarize the expedition's geological and glacial findings.

Ralph Tarr of Cornell University and Lawrence Martin of the University of Wisconsin participated in a series of explorations between 1905 and 1913 funded by USGS, the American Geographical Society, and NGS. O.D. von Engeln, a professional photographer and later professor of geomorphology at Cornell University, accompanied them several times. The Tarr and Martin surveys led to more than 50 scientific publications culminating in 1914 in *Alaska Glacier Studies of the National Geographic Society in the Yakutat Bay, Prince William Sound and Lower Copper River Regions*, one of the most comprehensive volumes ever produced on Alaska glaciers. Tarr and Martin's field investigations concentrated on the areas of Yakutat Bay, Malaspina Glacier, the Copper River, and Prince William Sound. In 1899, a tremendous earthquake with its epicenter near Yakutat had affected this entire region. Many of Tarr and Martin's early investigations centered around the earthquake's impact on the region's glaciers.

During the first decade of the twentieth century, the glaciers of Prince William Sound and the Kenai Peninsula were studied by U.S. Grant, D.F. Higgins, and Sidney Paige, of USGS. They investigated Valdez, Shoup, Columbia, Meares, and Barry Glaciers, as well as those of Port Nellie Juan, Icy Bay, Port Bainbridge, Blackstone Bay, and Thumb Cove of Resurrection Bay. They made notes, photographs, and maps of all the tidewater glaciers and many near tidewater. Their studies provide an excellent baseline for later comparisons.

> ... Surrounded by mountains of almost-silent floating ice in the fog and rain, I cannot quite account for the strange intensity of thought. I think the intensity and recognition go back. I think it has something to do with a man wrapped in animal hide and fur, crouched at the edge of a glacier twenty thousand years ago.
>
> (1965, Dave Bohn, from the anthology *A Republic of Rivers: Three Centuries of Nature Writing from Alaska and the Yukon*)

Following World War I, the number of investigations of Alaska glaciers significantly decreased, while the use of photography increased. Between 1926 and 1929, the U.S. Navy and USGS photographed many of Southeast's glaciers from the air. William Cooper of the University of Minnesota

Inside Kennicott Glacier, in Wrangell-St. Elias National Park, an icy passageway reveals the blue color of the ice. Notice the ice block partially hidden by sediment on the right. What looks like solid "ground" near a glacier may be debris-covered ice. (Fred Hirschmann)

investigated the flora and ecology of Glacier Bay. In 1935, he also visited Prince William Sound. These studies documented the cycle by which vegetation first becomes established on bare, recently deglaciated bedrock. He examined the succession of plants that follows, through the advent of a mature spruce and hemlock forest.

Expeditions in 1926 and 1929 produced the first aerial photographs of Alaska glaciers. Photos were obtained by four open-cockpit Loening amphibian biplanes, each equipped with a camera hatch in the bottom of the fuselage. In all, more than 30,000 aerial photographs were exposed with Bagley Tri-lens T-1 cameras. The 1926 expedition was so successful that a similar mission followed in 1929. Photographing glaciers was one of this expedition's highest priorities. Expedition leader R. H. Sargent, for whom the Sargent Icefield is named, wrote, "It was my purpose to have the fronts of all large glaciers in the region photographed, both by the mapping and the oblique cameras, so as to record the positions of the fronts of the glaciers in 1929. The oblique photographs taken from the side of the plane are marvels of grandeur and beauty, and the mapping photographs present wonderful views of the glaciers from directly overhead. These vertical photographs reveal the phenomenon of glacier flow in a manner never before recorded in the United States, so far as I know."

In 1926, William Osgood "Bill" Field Jr. made his first visit to Glacier Bay, one of many trips to Alaska, to relocate and occupy observation stations established by earlier observers, particularly those of Harry Fielding Reid. In 1931, Field visited Prince William Sound to reoccupy stations established by the Harriman Expedition in 1899. His photographic observations, designed to record the nature and positions of Alaska glaciers, continued into the early 1990s and resulted in numerous publications documenting

A man gazes over rugged landscape from his campsite above Herbert Glacier in Southeast's Tongass National Forest. Hiking trails lead from the highway north of Juneau to Herbert and nearby Eagle Glacier. (John Hyde)

Bradford Washburn: Photographer, Mountain Man

In the early 1930s, Bradford Washburn began photographic reconnaissance of Alaska mountains as a prelude to several of his climbing expeditions. Many of his photographs include the first aerial observations of glaciers in the Fairweather, St. Elias, Chugach, and Alaska Ranges. In 1933 and 1934, Washburn photographed the glaciers of the Fairweather Range between Finger and Malaspina Glaciers. In 1937, he photographed Lituya Bay and Fairweather Glacier, while in 1938, he photographed the area between Yakutat Bay and Bering Glacier. He is shown here holding a note he'd left near Mount McKinley in 1956 that describes his intentions for climbing Fake Peak, partly to recreate Dr. Frederick A. Cook's 1906 photo of McKinley's summit. (Photo by Brian Okonek)

RIGHT: *Austin Post relaxes on an Alaska hillside while conducting glacier research. Post grew up near Washington's Cascade Mountains and has studied mountain systems since he was a boy. (Bruce Molnia)*

FAR RIGHT: *Ascending a massive ice wall while wearing crampons and using an ice axe, a climber on Matanuska Glacier is surrounded by suncups, depressions in the glacier's surface created by warm, sunny conditions. (Tom Bol)*

glacier variations. In 1940, Field began working for the American Geographical Society, eventually establishing the World Data Center for Glaciology at its New York City location. There he began a project titled, "Observations of Glacier Behavior in Southern Alaska," which for more than 50 years systematically collected information about changes in Alaska's glaciers. Field died in 1994.

Following World War II, long-term field programs were established on the Juneau Icefield by Field and Maynard Miller; on the Malaspina-Seward Glacier system by Robert Sharp; and at various places in Alaska by Ohio State University's Institute of Polar Studies, the Arctic Institute of North America, and the University of Alaska. Beginning in 1955, Richard C. Hubley of the University of Washington began systematic aerial reconnaissance of Cascade Mountain glaciers.

In 1960, Austin Post, sponsored by the University of Washington and funded by the National Science Foundation, expanded this technique to cover most glaciers of western North America. In 1964, Post joined USGS, where he worked until his retirement in 1984. Many of the most spectacular aerial photographs of Alaska glaciers have resulted from his work. Beginning in the early 1970s, many photo missions were conducted in cooperation with Robert Krimmel of USGS, who frequently piloted the aircraft. Since the early 1980s Krimmel has carefully documented the catastrophic retreat of Columbia Glacier using vertical aerial photography. Typically, each flight recorded vertical and oblique views. Even in retirement, Post has continued his investigations, focusing on Columbia and Bering Glaciers.

In 1970, Larry Mayo of Fairbanks began seasonal aerial studies to examine short-term glacier fluctuations, especially of Gulkana and Wolverine Glaciers. His work continued until his retirement in 1999.

The 1964 Great Alaska Earthquake generated abundant scientific studies, mostly aimed at discovering how avalanches and landslides that fell on glaciers would change rates of melting and flow. More than 20 scientific papers focused just on the gigantic landslide that tumbled onto the terminus region of Sherman Glacier.

Vegetative studies, similar to the work of William Cooper, but also including long-term climate investigations based on the analysis of pollen profiles augmented by radiocarbon dating, have been conducted by Calvin J. Heusser since the 1950s. His 1960 publication *Late-Pleistocene Environments of North Pacific North America* serves as a framework on which to construct the post-Pleistocene history of southern Alaska. For example, scientists now know from pollen studies that spruce tree species growing in the Interior 9,000 years ago spread northwest, west, and south, while Southeast's coastal forests derived from those of the Pacific Northwest.

Life on Alaska Glaciers

Alaska glaciers are not biological or ecological wastelands. In addition to scientists who may live for part of a year or year-round on a glacier, plants and animals also inhabit them. Several species of animals use glacier surfaces seasonally, to escape insects or as a transportation corridor. Seals frequently haul out on recently calved icebergs near tidewater glaciers.

Many glaciers with thick surface sediment covers support the same types of vegetation that grow on adjacent land areas. Bering, Malaspina, and Fairweather Glaciers, to name just a few, support mature forests on their surfaces. Mature spruce trees older than 100 years and with diameters of more than three feet grow on the surface of Malaspina Glacier. As the stagnant ice melts, many trees and the soil in which they are rooted slide into developing meltwater lakes. During the

1993-1995 surge of Bering Glacier, sediment accumulated on the ice surface; less than one year later, fireweed, alder, willow, and spruce seedlings sprouted there, rooted in the thin layer of dirt.

Red algae flourishes on many glacier surfaces. In summer, during long, sun-filled days, this algae grows directly on the surface of snow covering glacial ice. It can reproduce so rapidly that it blankets large areas, giving the snow a reddish appearance. Red algae is actually a red-pigmented green algae able to make its own food through photosynthesis. Meltwater streams draining red snow fields sometimes carry so much red algae that the water has a pinkish tinge. I observed this situation on several glaciers including Bering Glacier and Cathedral Glacier, a small valley glacier that drains the eastern Juneau Icefield.

Medial moraines support a variety of plants and animals. Lichens grow on exposed rock surfaces. They play an important role in determining the sequence of events in a glaciated area. Compound plants composed of algae and fungus in a symbiotic relationship, lichens usually grow on bare rock surfaces. Studies by Roland Beschel in the 1950s showed that lichens were one of the first organisms to become established following ice retreat and that by measuring diameters of the largest members of known species, a growth curve could be constructed for a given geographic area. One of the most effective relative-age dating techniques practiced today, especially in areas where no radiocarbon-datable trees exist, is lichenometry, the dating of glacial events by analysis of lichen growth rates. The three species most commonly used in Alaska are Rhizocarpon, Umbilicarbius, and Lecidea.

Insects, small mammals, and many species of birds live on medial moraines or on adjacent sediment deposits. Spiders are common on many medial moraine surfaces. During May 1995, part of Bering Glacier's surging terminus overrode Pointed Island and covered more than 1,000 gull nests, many with eggs. Three years later as the island's surface began to emerge from under the retreating ice margin, the bird population quickly recolonized its old nesting ground.

Another animal inhabitant of Alaska glaciers is the snow flea or glacier flea, a vegetarian insect that belongs to the order Collembola. The main ingredient of the snow flea's diet is conifer pollen, sometimes supplemented with red algae.

Three times in recent years I have encountered bears using a glacier's surface as

FACING PAGE: *Tourists get a close-up view of an iceberg near the tidewater face of Chenega Glacier in Nassau Fiord, southwest Prince William Sound. (Hugh S. Rose)*

RIGHT: *Bering Glacier supports an iceberg-filled lake, encircled by crevasses. Bering, with an area of 2,000 square miles, is the largest glacier in continental North America. (Bruce Molnia)*

a transportation corridor. In one case, a black bear crossed the entire width of Surprise Glacier about 500 feet above the terminus to get from one side to the other, climbing over boulders, moraines, and around open crevasses. The bear took this route instead of crossing a relatively featureless outwash plain located at the foot of the glacier. In another case, a brown bear walked across the medial moraine at the head of Bering Glacier's piedmont lobe. Aside from vegetation on several nearby nunataks, this bear was nearly 20 miles from the closest forest.

Perhaps the most unusual situation I witnessed was a grizzly bear on Miles Glacier. While flying from Cordova to Bering Glacier, we chose the shortest route, which followed the length of Miles Glacier. As soon as we flew over the terminus; we saw a set of large bear tracks going up the center of the glacier. We followed them for more than a dozen miles until we came upon a big grizzly heading up the snow-covered middle glacier. After circling it several times, we continued on our way. Our last glimpse was of the bear's continuing beeline up the center of the glacier.

Ice Worms

By Penny Rennick, Editor

Northern poet Robert Service memorialized the ice worm in one of his ballads, and reporter Stroller White spread the fame of the ice worm cocktail. Only in the far North can a tiny creature take such a shortcut to mythology. But far from being a figment of a poet's imagination, the tiny ice worm is firmly frozen in reality.

"Watch where you step," came the command. Our Forest Service guide was alerting us that we were in ice worm country, where one misstep would certainly kill. Our group had spread out over Byron Glacier in early evening, looking for the mythical creatures.

We had come to the glacier in the evening because that's prime time for ice worms. These tiny, segmented worms, one-fiftieth of an inch in diameter and relatives of the earthworm, hide from the sun during the day. Thirty-two degrees Fahrenheit is ideal for them; warmer temperatures and the worms become mush, colder temperatures and they freeze to death. Why they thrive at temperatures where most life would die remains a mystery. "There's probably no other worm in the world that can develop from an egg to an adult at 32 degrees Fahrenheit," says biologist Dan Shain, an assistant professor at Rutgers University in New Jersey and a worm specialist.

As temperatures cool, the worms emerge from cracks in the firn, or less commonly from the ice itself, to feed on algae and detritus collected on the glacier. Ice worms wiggle about using tiny bristles called setae. How they move is the easy part; why and where they move is a puzzle. How do they find each other for mating? How do they form different colonies? Shain has documented ice worms scooting across a glacier at about 10 feet per hour. Not bad for an animal that's about three-fourths of an inch long. Shain thinks ice worms rely on sunlight, temperature, gravity, chemical trails on the ice, and an internal circadian rhythm to direct their movement. However they do it,

Ice worms thrive in temperatures just above freezing. Byron Glacier, near Portage Glacier southeast of Anchorage, is a good place to hunt for these tiny creatures. (Harvey Bowers)

Shain has documented ice worm colonies on several glaciers in Southcentral Alaska and the species has a known range from Washington state to Alaska. But much remains to be discovered, including the finer points of their navigation system.

Jim McAllister, Rai Behnert, and Art Bloom relax on the Death Valley Branch of Norris Glacier, near Juneau. Safe travel on glaciers, even in relatively flat areas such as this section of Norris, requires teamwork and experience. Using skis is one way to distribute your weight over a broad expanse, decreasing the risk of falling into a hidden crevasse. (Bruce Baker)

Glacier Travel

Glacier travel is not for the uninitiated. It is dangerous, sometimes fatal, and is one of the primary activities that leads to hypothermia. Do not travel alone on a glacier; if you are inexperienced, do not travel on one at all. Glacier accidents take their toll on the experienced glacier traveler as well as the novice. Almost every year the lives of experienced climbers and hikers are lost to hypothermia, in avalanches, or in crevasse or icefall accidents. Many of these deaths occur as climbers traverse glaciers on their way to peaks they will attempt to conquer.

The experienced glacier traveler knows the risks involved in glacier travel. However, most neophytes do not. Several years ago, Alaska newspapers contained many articles about the disappearance of a man at Mendenhall Glacier whose lifelong goal had been to "climb a glacier." Although he had the necessary equipment, crampons and an ice-axe, his lack of experience cost him his life. His body has not been found. Recently, a young camper fell into a crevasse while obtaining water. His body has also not been found. Crevasses often exist underneath covers of snow where they cannot be seen by glacier travelers. Many lives have been lost by glacier travelers falling into these covered crevasses.

If you must travel in a glacier environment keep the following ten rules in mind:
1. Do not travel alone in a glacial environment.
2. Always travel with an experienced companion.
3. Know the symptoms and treatment of hypothermia.
4. In crevassed areas, always rope-up and always probe for snow bridges.
5. Avoid icefalls, *séracs*, and avalanche areas.
6. Always carry emergency shelter, dry clothing, and extra food.
7. Know the fundamentals of using crampons, an ice-axe, and rope.
8. Whenever possible travel on skis or snowshoes as they distribute body weight over a much greater area than does travel by foot.
9. Do not travel in whiteout conditions. Make camp and wait for the weather to clear.
10. Use common sense and do not panic. ◻

Alaska's Glaciers: Area by Area

Alexander Archipelago

The Alexander Archipelago, the Southeast Alaska island group, extends about 300 miles from the Canadian border at Dixon Entrance to Cross Sound and Icy Strait. Small glaciers exist in mountainous areas on six of the islands: Revillagigedo, Prince of Wales, Kupreanof, Baranof, Chichagof, and Admiralty. Baranof has the most glaciers, with more than 60 miles of its crest covered by ice. None of the glaciers on any of these islands are named and none have been studied in detail. Examination of maps and aerial photography indicate that all show evidence of retreat and thinning.

Coast Mountains

The Coast Mountains straddle the U.S.-Canada border and form the mainland portion of Southeast Alaska's panhandle. They extend 425 miles from Portland Canal to Skagway. From east to west, the glacier-covered area of the Coast Mountains is more than 100 miles wide. This text discusses only the Alaska section.

Eighteen Coast Mountains glaciers are more than 10 miles long. Of these, four are entirely in Alaska and 14 cross the border, flowing from Canada to Alaska or vice versa. Included in the Alaska Coast Mountains are many ranges: Peabody Mountains, Rousseau Range, Halleck Range, Seward Mountains, Lincoln Mountains, Buddington Range, Kakuhan Range, Chilkoot Range, Sawtooth Range, and Takshanuk Mountains. All support glaciers. To the south, the glaciers are small and sparsely distributed. They increase in size and number to the north.

The greatest concentration of Coast Mountains glaciers are in two icefields, Stikine and Juneau. A number of smaller areas of ice accumulation straddle the crest of the Coast Mountains both to the north and southeast of the Stikine Icefield. South of the Stikine Icefield, Chickamin Glacier, northwest of Hyder, is the largest, at 50 square miles and more than 15 miles long. About half of the glacier is in Alaska. During the twentieth century, it retreated more than two miles. Soule Glacier, the second largest in the region, retreated about a mile during the second half of the twentieth century.

FACING PAGE: *Worthington Glacier, a valley glacier 29 road miles from Valdez, shows obvious evidence of retreat. The state recreation site here offers parking, a viewing shelter, interpretive displays, picnic sites, toilets, and a pay phone. (Bob Butterfield)*

RIGHT: *A Cessna 185 flies over an unnamed glacier east of Sitka on Baranof Island. Despite the scarcity of glaciers in the Alexander Archipelago, reminders of Pleistocene glaciation abound as U-shaped valleys, fiords, and other ice-carved rock. (R.E. Johnson)*

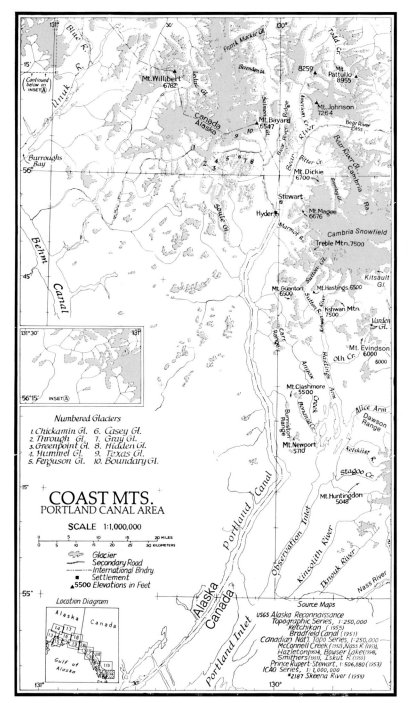

THE STIKINE ICEFIELD

The Stikine Icefield spans the crest of the Coast Mountains 120 miles from the Stikine River to the Whiting River and contains more than a dozen glaciers that are 10 or more miles long. Many of the larger glaciers descend from elevations of more than a mile and terminate in fiords of Frederick Sound or Stephens Passage.

LeConte Glacier is the southernmost active tidewater glacier in North America. More than 20 miles long, this prolific iceberg producer is popular with tourists. LeConte has retreated more than three miles since its position was first charted in 1887.

Flowing from the Stikine Icefield on the mainland between Wrangell and Petersburg, LeConte Glacier terminates in LeConte Bay and was named for a University of California geology professor. (Don Cornelius)

John Muir visited the glacier in 1879 and described it as one of the "most imposing" he had ever seen.

Baird and Patterson Glaciers are located in the Thomas Bay drainage. Advancing at the end of the nineteenth century, Patterson has retreated more than two miles during the past 75 years. Baird is unique in the Coast Mountains as it has a history of recent advance. Since the early nineteenth century when it began to form, a two-mile outwash plain has developed in front of its terminus.

Prior to that, Baird was a tidewater glacier.

Dawes and North Dawes Glaciers lie at the head of 30-mile Endicott Arm, a large fiord that drains into Stephens Passage. They have been studied since 1880 when they were first observed by Muir. Both are retreating rapidly. Twenty-three-mile-long Dawes is tidewater and has been retreating since the first mapping of its terminus. North Dawes was a calving glacier, but prior to 1923 it retreated out of the water and its terminus is now more than one-half mile from the shoreline.

Brown Glacier is a classic example of rapid retreat. In 1880, when Muir observed it, Brown had a tidewater terminus at the head of Fords Terror. He described majestic "shattered overleaning fragments" of ice falling from the tidewater terminus. By 1900, Brown had retreated about two miles. By 1950, retreat was more than four miles. Today, only small remnants composed of debris-covered melting-ice remain.

Dominating the head of Tracy Arm, a long sinuous fiord north of Endicott Arm, are Sawyer and South Sawyer Glaciers. About four miles apart, each stretches about 20 miles and covers more than 100 square miles. As recently as 1880, they may have been joined. The positions of both glaciers fluctuated during the twentieth century. However, both have retreated more than 25 miles from the time of their recent eighteenth-century maximum position. Presently, the terminus of Sawyer is stable, while South Sawyer appears to be retreating.

Between the Stikine and Juneau Icefields, Wright and Speel Glaciers are the only ones named. Wright, located south of the Taku River, has a length of 20 miles and an area of about 40 square miles. Since its first observation in 1891, Wright has retreated four miles.

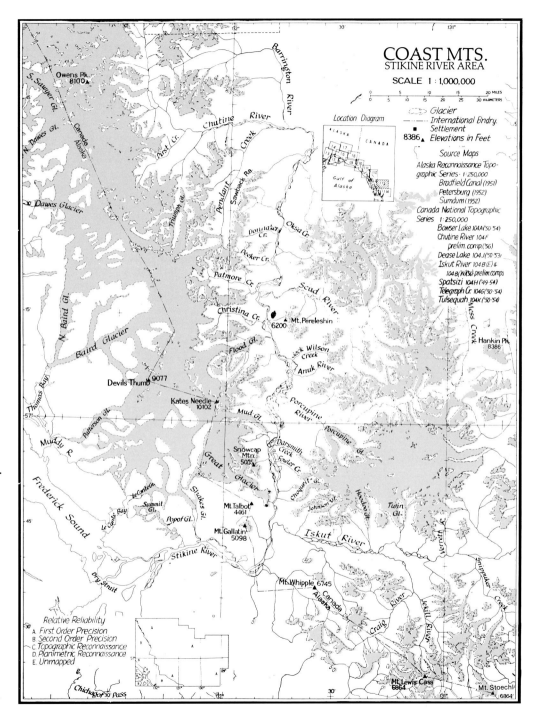

Sawyer Glacier's southern neighbor, South Sawyer, originating in the Coast Mountains east of Admiralty Island, produces some of the largest icebergs in Alaska. Boaters on Tracy Arm will likely encounter these mammoth chunks of ice. (Pieter Folkens)

THE JUNEAU ICEFIELD

The Juneau Icefield is the best-studied in Alaska and one of the most-studied in the world. It includes one of the most accessible glaciers in the state, Mendenhall. The Alaska section of the icefield covers more than 1,215 square miles with more than 30 valley glaciers that descend to near sea level.

Taku Glacier, at more than 30 miles long, is the largest that descends from the Juneau Icefield. John Muir said: "To see this one glacier is well worth a trip to Alaska." Since about 1890, its terminus has steadily advanced across Taku Inlet at a rate of several hundred feet per year. As it advanced, it knocked down trees in forests along its margins and overrode the moraines and outwash plain of Norris Glacier, immediately to its south. By the 1990s, its position stabilized. About 200 years ago, Taku Glacier advanced across Taku Inlet in a similar manner, forming a large lake that backed up into Canada. Prior to the mid-1930s, Taku Glacier ended in a deep fiord and calved icebergs into Taku Inlet. By 1939, sediment filled the inlet to the point that cruise ships, which previously had sailed right up to the glacier's terminus, could not enter the inlet. Taku's current terminus sits on a push moraine that restricts calving. This reduction in ice loss by calving is an important factor in reducing the retreat rate of calving glaciers. Geophysical studies determined that Taku has a maximum thickness of about one mile and reaches nearly 2,000 feet below sea level.

Hole-in-the-Wall Glacier is a **distributary** arm of Taku Glacier; about 1940, it thickened and expanded over the valley wall that previously contained it, advancing to the level of the Taku River. Taku and Hole-in-the-Wall Glaciers have created insurmountable problems for the construction of a road connecting Juneau to the Alaska Highway via the Taku River Valley. Consequently, Juneau is the only state capital in the United States not accessible by road.

Norris Glacier has a different advance and retreat history than Taku. Norris reached its most recent maximum position about 1917. Since then, it has retreated more than a mile and thinned significantly. For more than a century, its Dead Branch, once an active, ice-supplying tributary, has been a stagnant distributary arm that drains ice from the main glacier. Taku has a larger and higher area of accumulation than Norris. In fact, 40 percent of Taku's accumulation area is above

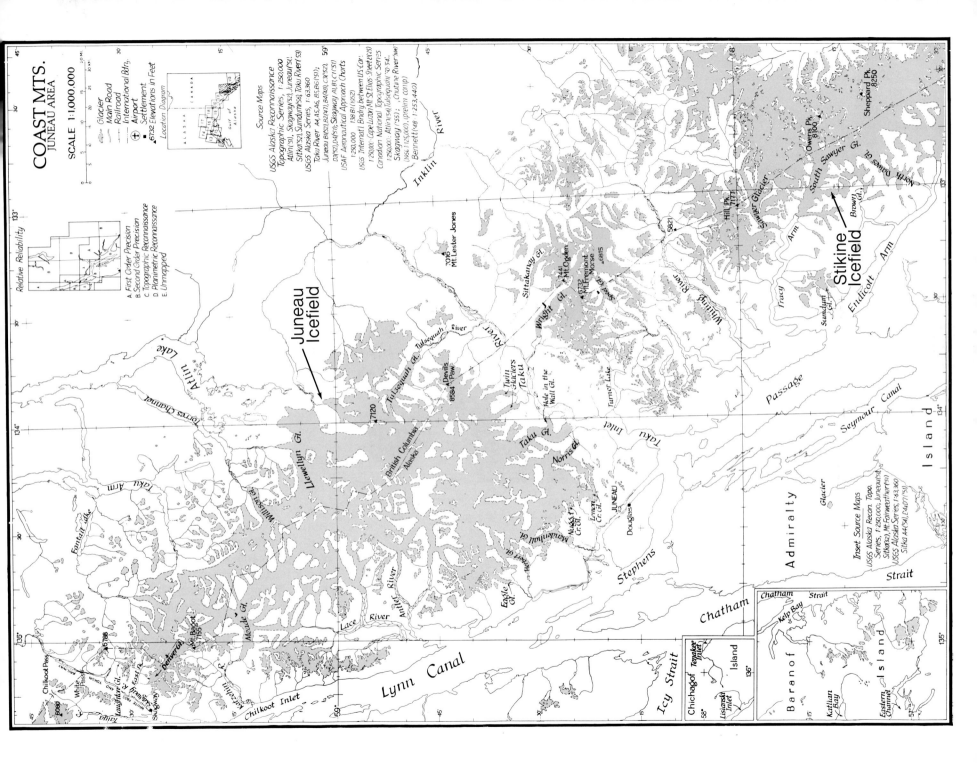

4,500 feet; only two percent of Norris's is that high.

Lemon Creek Glacier, one of the most-studied small glaciers of the Juneau Icefield, can be seen high above Gastineau Channel at the head of Lemon Creek Valley. It has retreated more than one and one-half miles during the last 250 years.

Less than 15 miles from downtown Juneau lies Mendenhall Glacier, 12 miles long. In 1879, Muir described it as one of the most beautiful of all the coast glaciers that are in the first stage of decadence. Its terminus, one of the best sources of large ice crystals for laboratory research, ends in Mendenhall Lake, which has a maximum depth of about 200 feet. Prior to the 1930s, most of the lake did not exist, as its basin was filled by the glacier. A U.S. Forest Service Visitor Center stands on a spot that was ice-covered as recently as 1940. Since then, Mendenhall Glacier has retreated about a mile and is now separated from the Visitor Center by the lake. Today Mendenhall is retreating several hundred feet per year. Thirteen distinct recessional moraines are located between its eighteenth-century terminal moraine and the late-1940's ice position.

Herbert and Eagle Glaciers, both north of Juneau, have similar retreat histories to that of Lemon Creek Glacier. Herbert has retreated an average of 60 feet per year since 1766, with a maximum average rate of retreat of 190 feet per year between 1928 and 1948.

Mead and Denver Glaciers are located east of Skagway. Mead is about 23 miles long, while Denver is much shorter. Both experienced significant retreat during the twentieth century. A century ago, Denver Glacier was one of the most-visited glaciers in Alaska.

North of Lynn Canal and east of the Chilkat River, a number of small- and medium-sized glaciers occur. Only three, Ferebee, Irene, and Chilkat, are named. Chilkat, with a length of 14 miles, is the largest. All of the glaciers in this region show evidence of recent retreat.

St. Elias Mountains

The St. Elias Mountains, 450 miles by 115 miles, straddle the Alaska-Canada border, paralleling the coastline of the northern Gulf of Alaska. The highest summit, Mount Logan (19,644 feet), is entirely within Canada, but high peaks that bridge the International Boundary or are in Alaska include Mounts St. Elias (18,008 feet), Bona (16,421 feet), Vancouver (15,700 feet), Fairweather (15,300 feet), Hubbard (14,950 feet), and more than two dozen other peaks with elevations greater than 10,000 feet. The St. Elias Mountains contain 50 glaciers with lengths greater than five miles. Thirty-seven of these are either partly or entirely in Alaska.

The Juneau Icefield Research Program

The Juneau Icefield Research Program (JIRP), organized in 1946, is the longest-running, continuous research program of an icefield system in the world. It was formulated to pursue long-term, multidisciplinary field research on topics related to arctic and mountain science. Among the disciplines investigated are the mechanics of glacier formation and movement, climatology, regional geology, biology and botany, and polar medicine. The program was originally established with support from the Office of Naval Research, in cooperation with the American Geographical Society and the U.S. Forest Service. In 1959, the Summer Institute of Glaciological and Arctic Sciences was organized to provide combined academic and field training for college and university students. This concept was expanded in the 1970s to include gifted high school students. The program is affiliated with the Universities of Alaska and Idaho.

Many leading professional scientists in the fields of glaciology, glacial geology, and arctic sciences have received their first glacial field experience and other important parts of their training through JIRP. The program is under the expert direction and leadership of Maynard M. Miller. My first experience with glaciers was as a student on the Juneau Icefield in 1968. I have returned many times as an instructor.

— Bruce Molnia

Researchers dig a rectangular snow pit to sample individual snow, firn, and ice layers within Taku Glacier, on the Juneau Icefield. Samples of known volume are weighed with a triple-beam balance to determine their density. This is one way to monitor the formation of glacier ice. (Bruce Molnia)

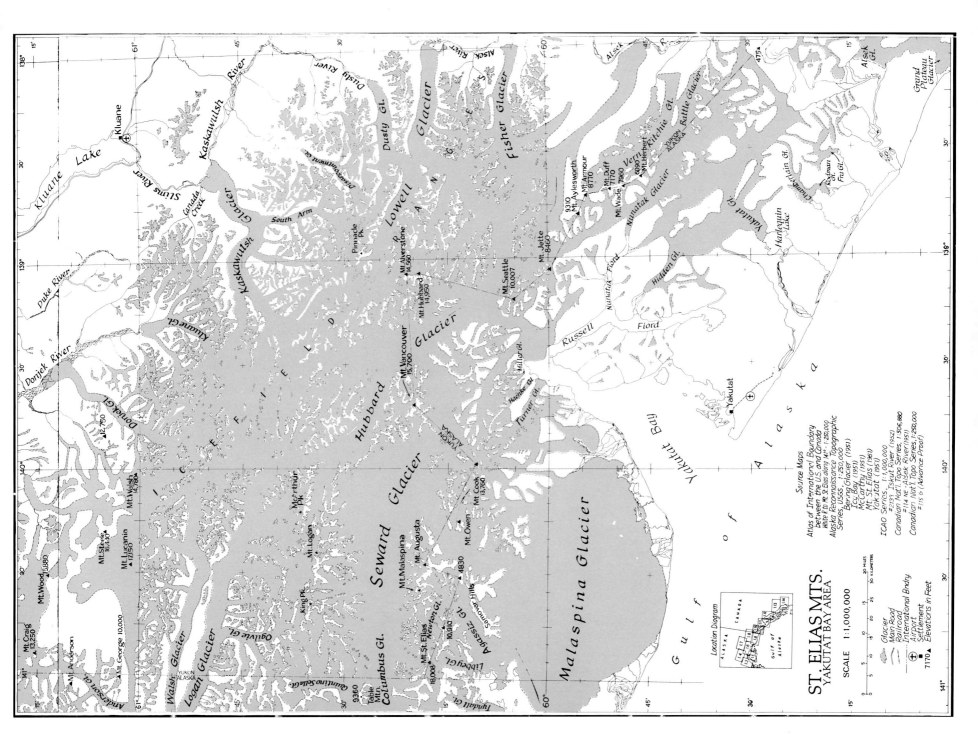

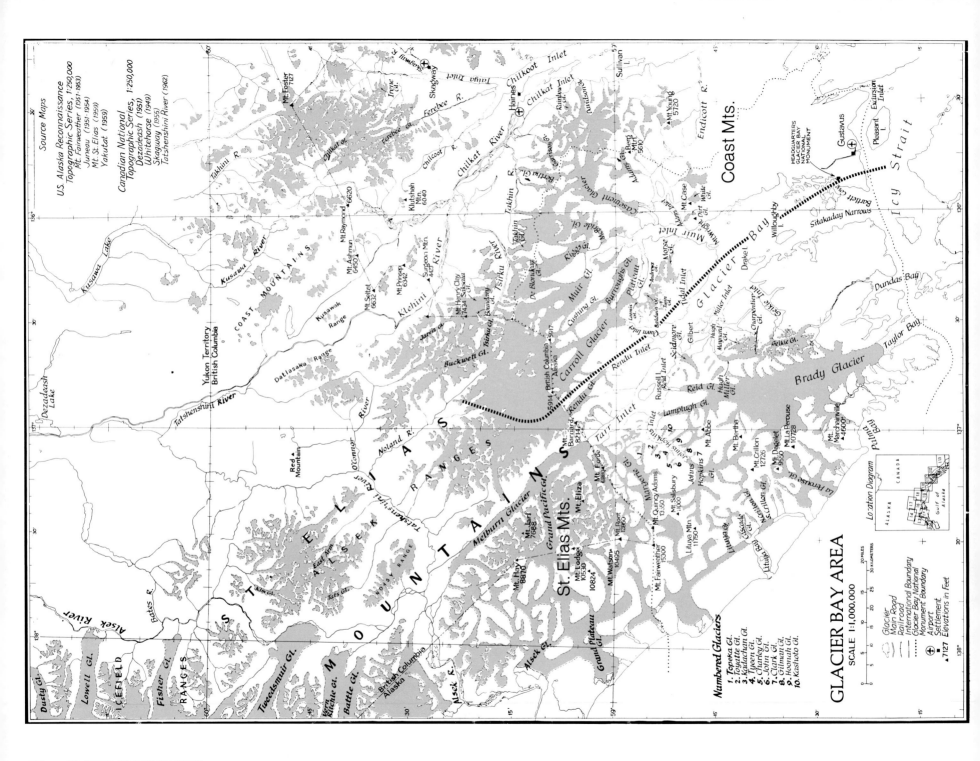

GLACIER BAY

Glacier Bay contains 12 active, calving, tidewater glaciers within its 65-mile length. It has the best-studied, most completely documented history of recent glacier fluctuations in Alaska. About 200 years ago, Glacier Bay did not exist. A single, gigantic ice mass extended into Icy Strait and completely filled the bay. By 1794, when Vancouver's lieutenant James Whidbey explored Icy Strait, the ice had retreated about six miles and had opened a small bay, the one shown on Vancouver's 1798 map of the northwest coast. In 1879 when Muir explored the bay, the ice retreat had exceeded 40 miles. By the end of the twentieth century, ice retreat exceeded 60 miles. Although most glaciers are still retreating, several have readvanced. At least 10 glaciers within Glacier Bay are greater than 10 miles, the longest being Grand Pacific.

Fiords of Glacier Bay that have actively calving, tidewater glaciers are Johns Hopkins, Tarr, Reid, and Muir Inlets. Muir, Rendu, Queen, Adams, Hugh Miller, Wachusett, and Geikie Inlets all had tidewater glaciers during the twentieth century that have since retreated onto land.

Johns Hopkins Inlet: In the 10 miles between Lamplugh Glacier at its mouth and Johns Hopkins Glacier at its head, nine glaciers descend the walls of this fiord. These range from small hanging glaciers such as Charley, John, and Clark to the tidewater termini of Toyatte, Kashoto, Hoonah, and Gilman. At the start of the twentieth century, glacier ice completely filled Johns Hopkins Inlet. Between 1892 and 1929 the glaciers retreated more than 11 miles. During most of the second half of the twentieth century, Johns Hopkins Glacier advanced more than two miles and thickened. Recently, a small retreat has occurred.

Tarr Inlet: Located at the head of Tarr Inlet are tidewater glaciers Margerie and Grand Pacific. The two were joined until 1912. Continued retreat of Grand Pacific brought its terminus north of the International Boundary into Canada sometime between 1913 and 1916. It remained in Canada until 1948 when it straddled the border. Beginning in 1961, Grand Pacific slowly readvanced to a position about one mile into Alaska. After joining Margerie in the 1980s, the two glaciers are now separated again. Geographically, Grand Pacific is unique in that it begins in the United States, extends through Canada for most of its length, and terminates in Alaska. No other tidewater glacier in Alaska leapfrogs across the International Boundary.

Caught in a cycle of advance and retreat, Glacier Bay's Grand Pacific Glacier (on the right) has a history of convergence with Margerie Glacier. This 1995 image shows their termini barely touching, but Grand Pacific has since retreated. Thousands of people visit Glacier Bay by cruise ship every summer to see the area's glaciers. (Harry M. Walker)

Twenty-five years ago, McBride Glacier's terminus reached as far as the sandy deltas shown in the foreground. Today, this Muir Inlet glacier is retreating, leaving behind scoured valley walls. A *tarn* nestles at the top of the peak on the glacier's inside curve. (Fred Hirschmann)

Muir Inlet: For many years, Muir Glacier was the most dynamic calving glacier in Glacier Bay. During the second half of the twentieth century, its principal tributaries Riggs and McBride Glaciers separated from its retreating terminus. By the end of the twentieth century, Muir's terminus had retreated above sea level.

FAIRWEATHER RANGE

The Fairweather Range, topped by 15,300-foot Mount Fairweather, runs 70 miles from

Grand Plateau and Grand Pacific Glaciers southeast to Cross Sound and Icy Strait.

Brady Glacier, with an area of 175 square miles, reached tidewater at the time of Vancouver's expedition. Through the late twentieth century, Brady advanced over four miles and developed a large outwash plain in front of its terminus. Today, Brady's terminus is stable, just behind its twentieth-century maximum position. However, its east and west margins show significant evidence of thinning.

La Perouse Glacier heads on the flanks of Mount La Perouse and Mount Dagelet and descends to sea level, where its three-mile-wide terminus is frequently washed by high tide and storm waves. In recent years the position of the ice terminus has fluctuated a few hundred feet but has always ended at the surf zone. In 1966 La Perouse Glacier made a mini-surge and ended in an ice cliff easily 300 feet high at the low tide line.

Lituya Bay, an I-shaped fiord, has three glaciers, Lituya, North Crillon, and Cascade, at its head. When La Perouse mapped the bay in 1786, it was T-shaped, with a pair of separate glaciers in each of the upper arms of the T. In each arm, the pair of glaciers joined and advanced. Together, North Crillon Glacier in the east arm of the T and Lituya Glacier in the west arm advanced about seven miles since 1786. Both have built large outwash plains that protect the glaciers from calving. In 1958, a gigantic, earthquake-generated rockslide fell into Gilbert Inlet, the west arm of the bay, and onto Lituya Glacier, shattering much of the terminus. The resulting splash wave removed soil and vegetation from surrounding slopes, up to an elevation of more than 1,750 feet.

NORTH OF THE FAIRWEATHER RANGE

Yakutat Bay is about 30 miles long and as much as 18 miles wide. Russell Fiord joins it at Disenchantment Bay, the narrow, northern neck of upper Yakutat Bay. Here, Hubbard, an immense valley glacier, reaches tidewater. Since 1894, when systematic mapping of Hubbard began, this glacier has been slowly advancing. Between 1895 and early 1986, Hubbard's advance narrowed Russell Fiord's entrance from 2.2 miles to less than 0.1 miles. From May to October 1986, the glacier's terminus sealed the fiord's mouth. Between May 29 and October 8, 1986, the ice dam maintained contact with the bedrock at Gilbert Point and the water level in Russell Lake rose 83 feet. Calving of icebergs from both sides of the 1,400-foot-wide ice dam into Russell Lake and Disenchantment Bay reduced the width of the dam until it failed. Within 24 hours, the water level in Russell Lake dropped more than 80 feet. The outburst that resulted was the greatest short-lived discharge of water in North America since glacial-lake outbursts occurred at the end of the Pleistocene Epoch (about 10,000 years ago). Today, Hubbard is the longest valley glacier in North America with a total length, including its Canadian source, of 92

Gulf of Alaska surf batters La Perouse Glacier's terminus with each high tide. In 1786, French explorer Jean Francois de Galaup de La Perouse lost two boats and 21 men when members of his crew ventured too near the entrance to Lituya Bay, 15 miles up the coast, and were drowned. (Bruce Molnia)

miles. Hubbard's tidewater terminus extends more than six miles and reaches heights of close to 300 feet. It is once again advancing toward Russell Fiord.

Variegated Glacier, a surging glacier, is about 15 miles long. At least six times in the last 100 years, it has experienced significant surges. The last surge in 1994 and 1995 resulted in a small ice advance. Previous surges were in 1905-1906, between 1911 and 1933, just prior to 1948, from 1964 to 1965, and between 1982 and 1983.

Turner Glacier, located on the west side of Disenchantment Bay, underwent catastrophic calving as the result of the July 9, 1958 earthquake, which had its epicenter near Lituya Bay. As much as 2,300 feet of Turner's tidewater terminus calved from the main body of the glacier. Today, this glacier is slowly advancing.

Many other carefully studied glaciers exist in the Disenchantment Bay-Russell Fiord area including Haenke, Miller, Hidden, and Nunatak Glaciers. All have been observed since the time of I.C. Russell in the early 1890s and by subsequent expeditions. All these glaciers are retreating.

Malaspina Glacier: Malaspina Glacier has the largest piedmont lobe in North America, with 850 square miles of ice at elevations of less than 2,000 feet. The lobe is about 45 miles from east to west and about 30 miles from north to south, with a circumference of almost 60 miles. More than 25 tributary glaciers supply Malaspina, the two largest being Seward and Agassiz. Malaspina Glacier has a total area, including its valley glacier section, of about 1,930 square miles; its size is frequently compared to that of Rhode Island. One of its most striking features is the contorted, folded, twisted moraines on its surface, produced by surges, the most recent of which occurred in 1986.

Piedmont lobe ice of Malaspina attains a thickness of between 1,130 and 2,050 feet and extends as much as 700 feet below sea level. Malaspina advanced to a maximum position in the last two centuries, but is now rapidly thinning and retreating. Melting removes approximately one cubic mile of water per year.

Icy Bay: In 1794, when Vancouver explored the Gulf of Alaska coast, Guyot Glacier extended into the gulf, filling the basin of Icy Bay. A separate bay existed about five miles to the east. By 1837, when Sir Edward Belcher sailed along the coast, sediment from Guyot had filled in the second bay. Guyot had also retreated, opening a small indentation at the mouth of the present bay. By 1886, Guyot had readvanced and ended about six miles seaward of the present-day

When the ice dam created by Hubbard Glacier's advance in 1986 finally weakened and broke, the resulting jokulhlaup *plucked deep-water shrimp off the bottom of Russell Fiord and deposited them on shore. (Bruce Molnia)*

How much does Malaspina Glacier weigh? Of course, no one knows, but scientists have demonstrated that some glaciers are heavy enough to cause Earth's crust to sink, a state called isostatic depression. Studies show that if these glaciers retreat, isostatic rebound occurs, and the land rises measurably. (Bruce Molnia)

shoreline. A large terminal moraine marks this maximum advance. Local legend tells of a Native village located on the west side of Icy Bay that was destroyed by the advance of a glacier. Retreat began as recently as 1904. To date, retreat exceeds 25 miles. Today, four separate fiords are located at the head of Icy Bay. Two are filled by arms of Guyot Glacier, one by Yahtse Glacier, and one by Tyndall Glacier. In early 2001, all of the fiords are lengthening as active ice retreats and calving continues.

Walsh Glacier, located astride the International Boundary 60 miles north of Icy Bay, is about 35 miles long and covers an area of 160 square miles. During the period 1960-1966, Walsh Glacier underwent a surge that resulted in ice movement of about 10 miles, with a maximum advance of two miles between August 1965 and September 1966.

Barnard Glacier, not to be confused with Barnard Glacier in College Fiord, is about 30 miles long with its terminus just north of Chitina River. In 1938, Bradford Washburn made a spectacular photograph of Barnard Glacier that showed numerous straight, parallel, medial moraines, many unwavering for more than five miles. It is this photograph that has brought Barnard Glacier its fame.

The Skolai Mountains contain more than a dozen glaciers with lengths of at least five miles. The largest, Russell Glacier, is 27 miles long. Its most recent advance occurred between 1881 and 1921. Twisted and bent pieces of an old horse carriage were found buried in its terminal moraine. All Skolai Mountains glaciers show evidence of continued retreat.

Wrangell Mountains

The Wrangell Mountains, a large, young, volcanic massif 100 miles long and 60 miles wide, are located between Nabesna (to the north), the Skolai Mountains of the St. Elias Mountains (to the east), Chitina River (to the south) and Copper River (to the west). Mounts Blackburn (16,390 feet), Sanford (16,237 feet), Wrangell (14,163 feet), and Drum (12,010 feet) are the principal peaks. All are either active or dormant volcanoes or are composed of layered volcanic material. Glaciers of the Wrangell Mountains cover an area of more than 3,200 square miles and include a broad upland icefield that extends from Mount Blackburn to Mount Sanford, a distance of more than 25 miles.

Among the valley glaciers that drain this area are Nabesna, the largest inland glacier in

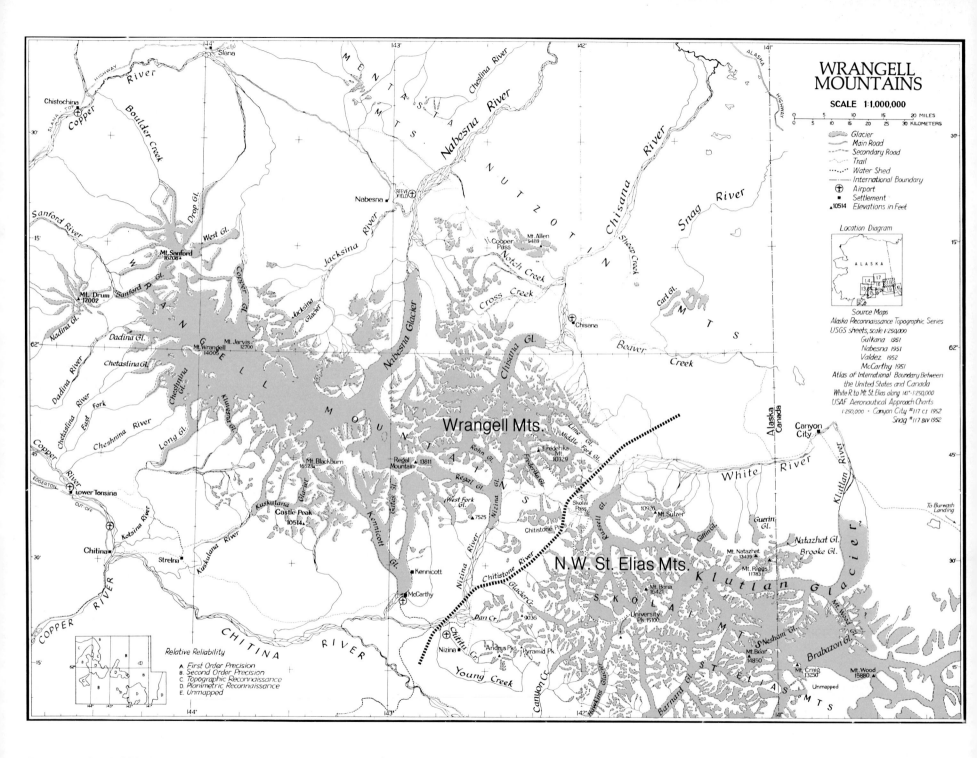

North America, and Copper, Sanford, Cheshnina, Long, and Kuskulana. Other large glaciers that descend from snowfields in the eastern Wrangell Mountains include Chisana, Nizina, and Kennicott.

Mount Wrangell is the only currently active volcano in the Wrangell Mountains. Studies show that Mount Wrangell's heat flow increased markedly during the second half of the twentieth century. Glaciological-volcanological research conducted by the University of Alaska Fairbanks has attempted to measure the effect of increasing volcanic heat flow on Mount Wrangell's summit glaciers. Changes occurring there have been spectacular. About 100 million cubic yards of ice has melted from the interior and immediate surrounding area of one of the three active craters along the rim of the volcano's summit caldera. The melting rate of glacier ice is being measured annually as a means of estimating the volcanic heat flow. Thus, glaciological research is being used as a tool in volcanological studies. Here, Mount Wrangell's glaciers are being used as a huge natural "calorimeter."

Many other glaciers in the Wrangell Mountains have been investigated in the past. Nabesna Glacier, with a length of about 50 miles and an area of nearly 400 square miles, may have as many as 40 tributaries. In spite of its enormity, examination of aerial photographs of the glacier suggests that Nabesna is slowly retreating.

Chisana Glacier, second-largest glacier in the Wrangell Mountains at almost 30 miles long and with an area of about 140 square miles, retreated during the first half of the twentieth century. Since the 1950s, Chisana's terminus has fluctuated and the glacier has experienced a number of small surges. Three or four recent moraines are located adjacent to the active ice margin.

Chugach Mountains

The Chugach Mountains, which form an arc around the northern Gulf of Alaska, contain about one-third of Alaska's glacier-covered area, including the largest glacier in continental North America: Bering Glacier. In addition to its piedmont lobe, Bering has a valley glacier section more than 100 miles in length, formally called the Bagley Icevalley. Bering's piedmont lobe is 25 miles across in both east-west and north-south directions. The entire glacier has an area of 2,000 square miles, 70 square miles larger than Malaspina. Bering's primary western tributary is Steller Glacier. Twentieth-century retreat has resulted in development of a series of ice-

Meltwater from Hole-in-the-Wall Glacier in the Wrangell Mountains runs into Skolai Creek, then flows down the Nizina, Chitina, and Copper Rivers to the sea. This view is from Chitistone Pass. (Curvin Metzler)

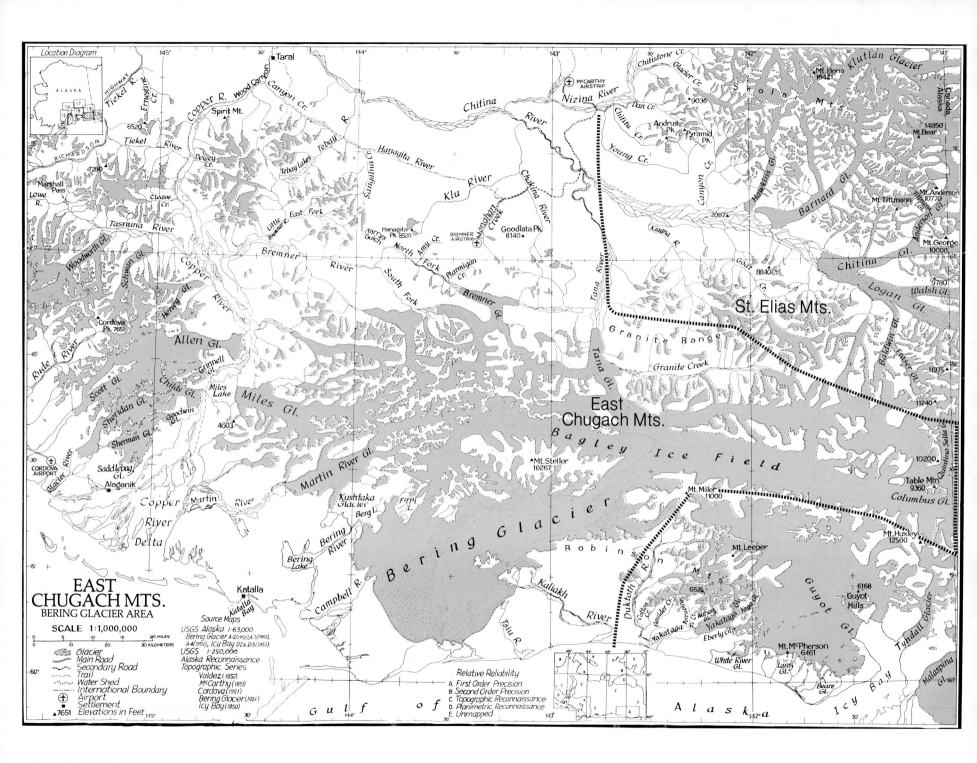

marginal lakes along much of Bering's terminus. Large icebergs calve directly into these lakes. Between 1993 and 1995, Bering Glacier surged six miles. In all, six surges occurred during the twentieth century.

Part of Bering Glacier's terminus is covered by a mature forest, growing on debris-covered stagnant ice. Much of the terminus region of 30-mile Martin River Glacier is also covered by ablation moraine and a number of small lakes. Like nearby Bering, it is stagnating, **downwasting**, and slowly retreating.

Copper River Region: A number of large glaciers make up part of the shoreline of the lower Copper River or drain into it or its delta. A braided stream, the Copper is the largest river draining into the Gulf of Alaska. It carries more sediment than any river in Alaska, including the Yukon. In 1905, when large quantities of copper ore were discovered in the Wrangell Mountains, a route to transport the ore to the port of Cordova was needed. The route selected along the Copper River had to pass over a stretch that contained Heney, Allen, Childs, Grinnell, Miles, and Goodwin Glaciers. Between 1906 and 1910, the Copper River and Northwestern Railroad was constructed at a cost of about 20 million dollars. Tracks had to be laid on five miles of Allen Glacier, one-quarter mile of moraine-covered ice of Grinnell Glacier, and about one-half mile of terminal moraine of Heney Glacier. Through the 1930s, work crews continuously struggled to maintain the track and compensate for glacier movements and floods.

Miles Glacier forms the eastern bank of the Copper River at Miles Lake. Icebergs calved from the glacier float under the Million Dollar Bridge. Massive steel and concrete iceberg deflectors have protected the bridge since it was built in 1909. In 1888, Miles's terminus was only 400 feet from the bridge site. Following a decade of retreat, it rapidly advanced toward the bridge during construction, stopping less than 1,000 feet away. Since then, Miles Glacier has retreated several miles. Its detached, stagnant-ice terminus is located on the west side of the Copper River.

Twelve miles long, Childs Glacier forms the west shoreline of the Copper River for four miles. During bridge construction, Childs Glacier advanced about 2,000 feet, stopping only 1,474 feet from the bridge.

Shaken loose from Shattered Peak during the 1964 Good Friday earthquake, a huge rockslide covers Sherman Glacier. USGS geologist Lawrence Martin named the glacier in 1910 in honor of Civil War General William Tecumseh Sherman. Martin's notes read, "he (Sherman) said 'war is hell;' so I put him on ice." (Bruce Molnia)

The advance ended in 1912 and Childs has not threatened the bridge since.

Seven miles north, the detached, stagnant-ice, former terminus of 20-mile Allen Glacier flanks the west bank of the Copper River. Track was laid on ice along the river's edge.

Glaciers along College Fiord's western wall are, from left to right, Wellesley, Vassar, Bryn Mawr, and Smith. Harvard, at the head of the fiord's Harvard Arm, is slowly advancing. John Muir wrote of this group, "they come bounding down a smooth mountain side through the midst of lush flowery gardens and goat pastures, like tremendous leaping, dancing cataracts in prime of flood." (Loren Taft)

Tarr and Martin observed that the ballast beneath the ties and rails of the railway actually rested upon the ice, not upon an abandoned moraine as at Heney Glacier. They wrote that problems encountered included settling of the track as ice beneath it melted and frequent breaking out of new streams. Ice movement shifted bridge supports by 18 inches. Advances of Allen and Grinnell Glaciers occurred, but had no effect on railroad operations. For more than 20 years, trains crossed the glaciers daily in both directions.

The March 27, 1964 Alaska Earthquake generated large rockslides and avalanches that fell on many Chugach Mountains glaciers. The largest was a gigantic landslide that tumbled onto Sherman Glacier, west of the Copper River. The slide covered three square miles with more than 25 million cubic yards of rock debris. All the rock was derived from one mountain, Shattered Peak. The debris moved on a cushion of air at up to 50 miles per hour, overriding a ridge about 500 feet high. The following year, the ice surface between the slide and the terminus was lowered between 26 and 33 feet, due to anomalous air temperatures that resulted from the heating of air as it passed over debris. Some scientists postulated that the debris would cause a surge of the glacier by 1978. This has not happened.

At least 25 other glaciers experienced earthquake-generated rockslides or avalanches, including Childs, Allen, Slide, Miles, Schwan, Martin River, Bering, Steller, Sheridan, Columbia, Serpentine, Meares, Smith, Harvard, Yale, Vassar, Surprise, and Harriman Glaciers. Causes other than the 1964 earthquake have also generated rockslides. At least 11 glaciers experienced rockslide avalanches between 1945 and 1963. The largest was on Barry Glacier in 1960.

Prince William Sound: Prince William Sound contains the greatest concentration of tidewater, calving glaciers in Alaska. Chugach and Kenai Mountains glaciers have cut more than 40 fiords into the margins of the sound, one-third of which contains its 20 active tidewater glaciers. Fiords with calving glaciers are located on the west and northwest side of the sound and include College, Harriman, and Nassau Fiords; Columbia, Blackstone, Shoup, and Icy Bays; Barry, Yale, and Harvard Arms; Unakwik Inlet; and Ports Wells, Bainbridge, and Nellie Juan. In addition to those in the fiords, glaciers exist on Montague and Knight Islands.

College Fiord, 25 miles long and three miles wide, contains six calving tidewater glaciers: Harvard and Yale Glaciers, in arms at the fiord's head, and Smith, Bryn Mawr, Vassar, and Wellesley Glaciers on its west wall. Named, non-tidewater glaciers in College Fiord are Barnard, Holyoke, Radcliffe, Amherst, Dartmouth, Lafayette, Baltimore, Eliot, and Crescent. Except for Crescent, these glaciers were named by members of the Harriman Alaska Expedition for colleges and universities with which they had affiliations. Included are all Seven Sisters colleges, several Ivy League and northeastern colleges, two "finishing schools," and two past presidents of Harvard University. Most of College Fiord's glaciers have shown a general trend of slow retreat since their initial description. Recently, Yale Glacier has retreated nearly four miles. Harvard Glacier, however, has been slowly advancing for much of the past 100 years.

In 1899, when the Harriman Alaska Expedition explored Barry Arm, Barry Glacier almost completely filled it. Upon orders of expedition leader Edward Harriman, Capt. Peter Doran sailed closer to the ice face than he thought prudent, and the expedition was able to skirt Barry Glacier and discover a completely unknown inlet, which they named Harriman Fiord. This 1899 position of Barry Glacier was the maximum reached during the past 500 years or more. Between 1899 and 1914, the glacier retreated more than four miles. Today, three glaciers are located at the head of Barry Arm: Barry, Cascade, and Coxe. Coxe Glacier separated from retreating Barry Glacier between 1910 and 1914. At the start of the twenty-first century, Barry and Cascade Glaciers were barely connected.

Harriman Fiord contains more than a dozen glaciers, the longest being Harriman Glacier at its head. All of the glaciers except Harriman have retreated since first seen by the Harriman Expedition. Harriman Glacier slowly advanced during the second half of the twentieth century and is presently slowly retreating. Surprise Glacier, named because it was the first glacier seen when the expedition entered Harriman Fiord, retreated about one and one-quarter miles between 1899 and 1910. Today it is still slowly retreating.

Meares Glacier, at the head of Unakwik Inlet, a 20-mile northward-trending fiord, was studied by U.S. Grant and D.F. Higgins in 1905 and 1909 and Lawrence Martin in 1910. Meares is an iceberg-calving, tidewater glacier that advanced for most of the twentieth century. Twice, 1931-1935 and 1966-1970, the terminus position remained stable. Total advance has been about one-half mile. In 2000, Meares Glacier's terminus was slowly advancing and pushing down trees, although it showed some evidence of thinning along its margins.

Named for Columbia University, Columbia Glacier is rapidly retreating and producing

The upper section of Surprise Glacier, which empties into Harriman Fiord, lies in the Municipality of Anchorage, Alaska's largest city. (Jon R. Nickles)

huge quantities of icebergs. Between 1910 and early 1979, part of Columbia's terminus grounded on Heather Island. But in January 1979, the glacier retreated off the island and began an irreversible, drastic retreat. It has since retreated more than six miles. The largest glacier in Prince William Sound, Columbia is 35 miles long and covers more than 430 square miles. Due to its proximity to the port of Valdez and the terminus of the trans-Alaska oil pipeline, tankers carrying petroleum from Valdez to refineries in the contiguous forty-eight states may be restricted in terms of when and where they can travel in Prince William Sound to minimize tanker-iceberg interactions. In 1979, USGS predicted that the retreat of Columbia was inevitable, that it "probably will begin in less than 20 years, and that iceberg production would increase to four times the number of icebergs produced in 1977-1978." They were correct on all accounts. Ultimately, a new 25-mile fiord will open.

Valdez Glacier's recent history is tied closely to the early-twentieth-century gold rush, when the glacier served as a major corridor from the port of Valdez to the gold fields of the Interior. An 18-mile route led from Valdez Glacier through the mountains to the head of Klutina Glacier. About 1910, mining began in bedrock exposed on both walls of Valdez Glacier's valley. Thinning of the glacier has left these claims more than 300 feet above the present ice surface. During the twentieth century, Valdez Glacier retreated about a mile.

NORTHERN CHUGACH MOUNTAINS

Twenty-eight-mile Matanuska Glacier, 80 miles northeast of Anchorage, is visible from the Glenn Highway. Several times during the twentieth century its terminus advanced and overrode stagnant ice and ablation moraine. During the summer of 1979, the terminus advanced more than 100 feet in 60 days. A moraine one-quarter mile in front of the terminus is less than 200 years old.

Worthington Glacier, a small mountain glacier, lies just north of Thompson Pass, about 20 miles north of Valdez. A new interpretive display has opened adjacent to its terminus, which is less than one-half mile from the Richardson Highway. The glacier shows conspicuous evidence of recent retreat as huge lateral and terminal moraines surround it. Worthington has thinned so much that visitors need to walk downhill from the parking lot to reach the glacier terminus.

Twenty-five-mile Knik Glacier has a history of blocking the flow of Lake George and Colony Glaciers from entering the Knik River. Often, meltwater would pond into a large lake, called Lake George, which covered an area of as much as 25 square

Icebergs cover the water at the terminus of Columbia Glacier, the largest in Prince William Sound. As Columbia rapidly retreats, a new fiord is opening. (Hugh S. Rose)

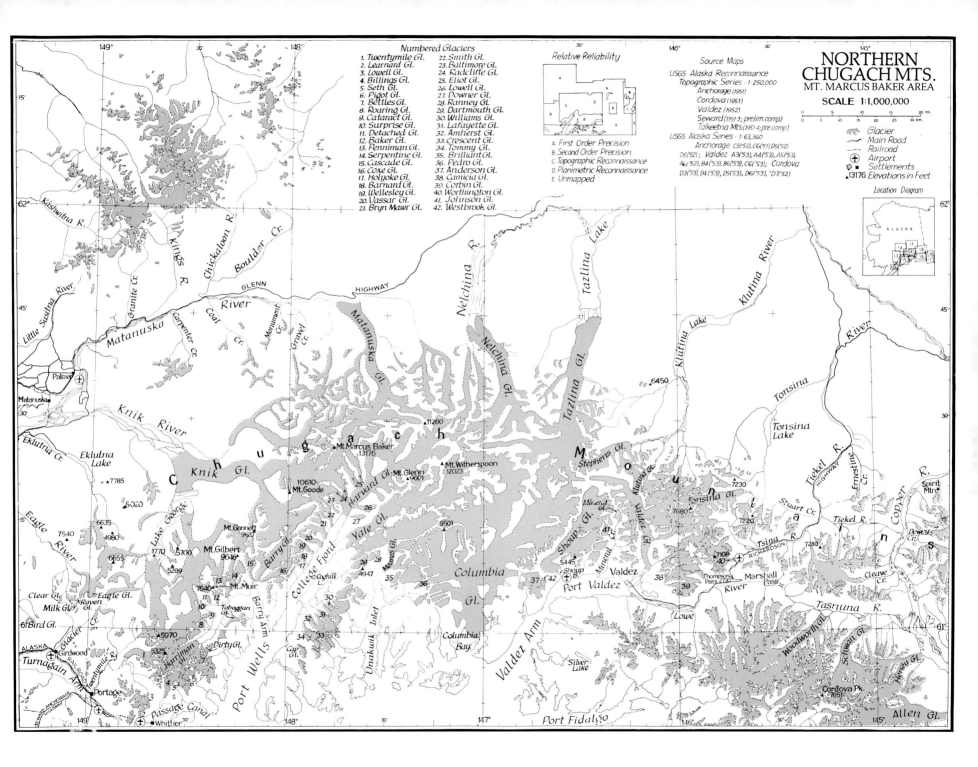

ALASKA GEOGRAPHIC® 79

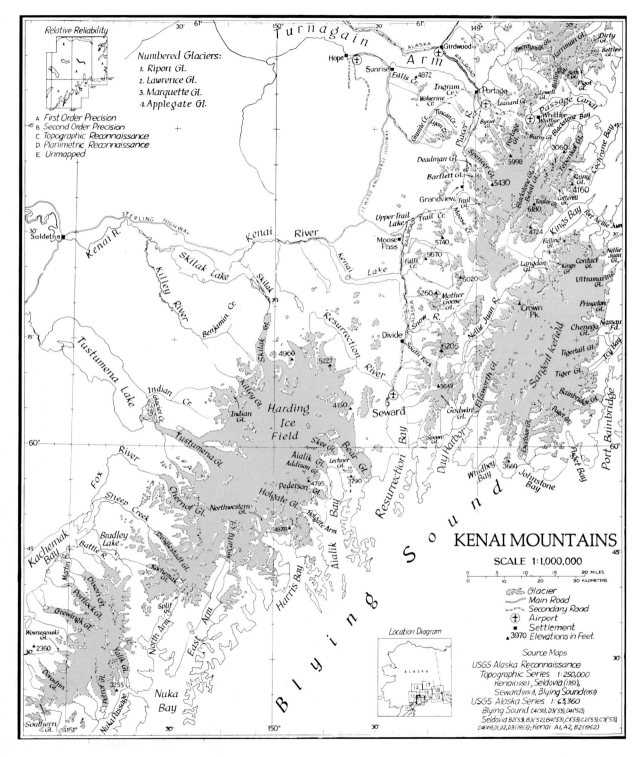

miles. Eventually, the lake would overtop its ice dam and begin a *jokulhlaup*, with a peak discharge of 150,000,000 gallons per minute, that would last for weeks and attract sightseers. Since 1966, Knik Glacier has failed to advance sufficiently to seal off the flood channel. In 2000, the river channel between the glacier and bedrock was only a few hundred feet wide. A minor readvance is all that would be needed to once again impound the meltwater.

Kenai Mountains

The Kenai Mountains are 120 miles long by 20 miles wide and extend the entire length of the Kenai Peninsula. Two large icefields, Sargent and Harding, and two smaller, unnamed icefields are located in the uplands. Combined, Harding and Sargent Icefields cover more than 2,000 square miles and have glaciers that descend into both Prince William Sound and Cook Inlet drainages, many reaching to near sea level. Seven glaciers have lengths exceeding 12 miles; two exceed 18 miles.

Eight glaciers descend into Blackstone Bay from an unnamed, 200-square-mile icefield. The area is named for a miner who lost his life on Blackstone Glacier in 1896. Like College Fiord, many of the glaciers here were named for universities. These include Ripon, Lawrence, Marquette, Beloit, and Northland Glaciers, all named in 1910 by Martin for schools in Wisconsin. Today, all of the glaciers in Blackstone Bay appear to be retreating. Other major glaciers that drain from this icefield include Whittier, Portage, Trail, Spencer, Skookum, Burns, Bartlett, Taylor, Tebenkof, and Wolverine; all have retreated significantly since first observed.

Before it retreated around a bend in its

valley, Portage Glacier was the most visited glacier in the Anchorage area and second-most visited glacier in Alaska. For centuries it served as an overland passage, a portage, between Turnagain Arm and Prince William Sound. About 1880, Portage Glacier stood at its recent maximum position. Since then, it has retreated more than two miles. The recently opened approach to the tunnel to Whittier provides a new vantage point from which to view the glacier. Continued retreat at the present rate would lengthen the lake by about one-half mile and put the terminus at a point of stability by 2020. Portage Glacier is known for the tremendous accumulations of icebergs that drift to the shoreline in front of the Begich, Boggs Visitor Center.

More than 25 glaciers drain the 35-mile by 20-mile Sargent Icefield. The largest of these are Chenega, Princeton, Excelsior, and Ellsworth. Ellsworth Glacier, more than 17 miles long, empties into a moraine-dammed lake above the head of Day Harbor. Chenega Glacier, more than a dozen miles long, has an area greater than 125 square miles. All these outlet glaciers show significant retreat and thinning.

The 50-mile by 30-mile Harding Icefield is the largest in the Kenai Mountains. Four glaciers with lengths greater than 15 miles descend from its summits, including Skilak and Tustumena, each about 20 miles long. Many others, including McCarty Glacier in McCarty Fiord, Northwestern and several unnamed glaciers in Northwestern Fiord, Holgate Glacier in Holgate Arm, and Aialik Glacier at the head of Aialik Bay have tidewater termini. All retreated significantly during the twentieth century. McCarty retreated nearly 20 miles, while Northwestern, named for Northwestern University, retreated nearly 10 miles. Many early-twentieth-century, former tributaries to the lower part of Northwestern Glacier are now small relict ice patches stranded high above Harris Bay. The largest of these, an unnamed glacier, had a tidewater terminus in 1950, but retreated to an elevation of about 1,000 feet by 1964. All Harding Icefield glaciers show evidence of recent thinning and retreat.

South of Harding Icefield and visible from Homer across Kachemak Bay is another unnamed icefield, informally called the Grewingk-Yalik Glacier Complex. This icefield is 26 miles across from east to west. Petrof and Yalik Glaciers are the major glaciers draining the east side of the complex, while Dixon, Portlock, Grewingk, Wosnesenski, and Doroshin Glaciers are the major glaciers that drain the west side. Grewingk is 13 miles long and covers about 30 square miles. Comparison of recent aerial photography with early observations shows that each of these glaciers substantially retreated during the twentieth century.

This beach below Excelsior Glacier can be reached by boat from Seward. (Bruce Molnia)

Sixty miles southeast of Alaska's mountaineering hub, Talkeetna, backpackers explore ablation moraine of Talkeetna Glacier. The glacier feeds what the Dena'ina tribe called the "river of plenty." (Brian McCullough)

Talkeetna Mountains

The 100-mile-long by 80-mile-wide Talkeetna Mountains, located between the Chugach Mountains and the Alaska Range, contain several hundred, generally unnamed, valley and cirque glaciers at elevations of 4,000 to 8,000 feet. Glaciers are distributed throughout much of the upland region of the mountains with the greatest concentration on the south side, in the mountains overlooking headwaters of the Talkeetna River. The largest glaciers are also located here; seven exceed five miles in length. The largest, more than 10 miles long, forms the headwaters of Sheep River and is unnamed. Talkeetna and Chickaloon Glaciers, both eight miles long, are the second and third largest. Their termini are covered with thick ablation moraines and show evidence of substantial retreat and thinning. Based on comparisons of maps and photography, one study showed that at least 50 glaciers east of the Talkeetna and Chickaloon Rivers either decreased significantly in area or completely melted away during the last half of the twentieth century. All of the Talkeetna Mountains glaciers that descend below an elevation of 5,000 feet show significant thinning and retreat.

Alaska Range

The Alaska Range, a 650-mile arcuate mountain system, extends from near the Canadian border to the south side of the Neacola Mountains at the head of the Alaska Peninsula. More than 5,000 square miles of the range is covered by glaciers that extend in elevation from more than 20,000 feet to less than 300 feet, from the summit slopes of Mount McKinley to the foreland of Cook Inlet. The Alaska Range comprises many mountain groups that support glaciers. From east to west, these are Nutzotin, Mentasta, Clearwater, Kichatna, Teocalli, Tordrillo, Terra Cotta, Revelation, and Neacola Mountains.

The Nutzotins may contain as many as 100 small cirque glaciers in two accumulation areas, around Mount Allen and north of Klein Creek, on either side of the Chisana River. Both areas are about 8,500 feet in elevation, their total area less than 50 square miles. Only one glacier, three-mile Carl Glacier, located 15 miles west of the Canadian border, is named. Alfred Hulse Brooks and Stephen Capps briefly investigated glaciers of this area early in the twentieth century, but these glaciers have not been studied since.

Several dozen small cirque glaciers flow from the Mentasta Mountains in an area about 15 miles north of Nabesna. The largest group is clustered around 8,235-foot Noyes Mountain. All are unnamed. Like the Nutzotin Mountains glaciers, these are virtually unknown.

Glaciers of the Mount Kimball (9,007 feet) area cover 325 square miles. Most descend from a 50-mile interconnected icefield that is located astride the crest of the eastern Alaska Range. At least 19 valley glaciers that head along the crest have lengths of five miles or

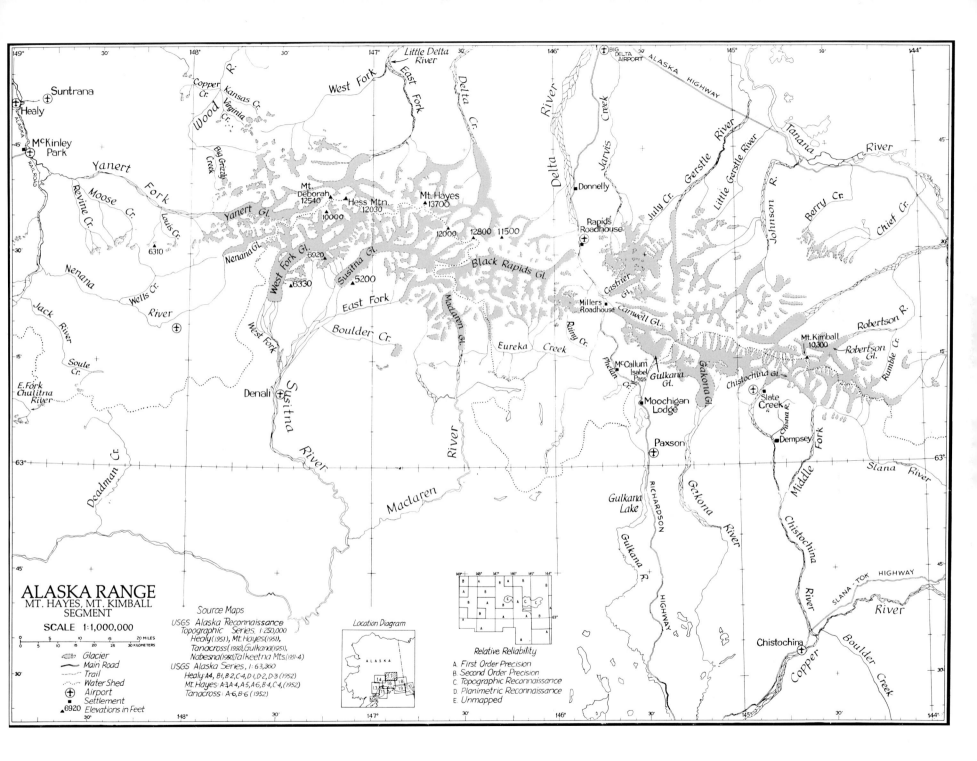

more; the largest, 75-square-mile, 18-mile-long Johnson Glacier, forms the headwaters of Johnson River. Except for Castner, Canwell, and Gulkana Glaciers, these glaciers are unknown. Canwell Glacier was first investigated in 1902 by W.C. Mendenhall and T.G. Gerdine. This 15-mile glacier and its neighbor, Castner Glacier, had advanced about a mile at the start of the Little Ice Age. Canwell Glacier had a second smaller advance prior to the twentieth century. A retreat of more than one mile followed. Between 1902 and 1941 Canwell Glacier again advanced. Photographs suggest that this advance was also about a mile. Both are now retreating.

Since the early 1960s Gulkana Glacier, four miles east of the Richardson Highway, has been studied intensively by the University of Alaska and USGS to determine geophysical parameters of the glacier, foliation patterns, structure, flow patterns, and mass balance. One study determined the configuration of Gulkana's valley with a gravimeter survey from the ice surface. It discovered that Gulkana Glacier's valley consists of two parallel bedrock channels, separated by a medial ridge. Ice in the eastern channel was 750 feet thick, while in the shallower, western valley the ice was only 430 feet thick. Gulkana Glacier, which has a length of about seven miles, was photographed by F.H. Moffit in 1910. Photographs made in 1960 from Moffit's photo site and photographs made in subsequent years show that the terminus has retreated several miles. Since 1966, annual mass balance studies of Gulkana Glacier have been conducted.

Jarvis Glacier, like Gulkana Glacier, was studied using gravimetric techniques. This glacier, which lies just north of Gulkana Glacier, is only five miles long. The gravity survey showed that Jarvis Glacier lies in a deep U-shaped valley with ice thickness of more than 1,050 feet.

Mount Hayes (13,832 feet) is the highest peak in the eastcentral Alaska Range. Together with Mounts Moffit (13,020 feet) and Deborah (12,339 feet), the region has a glacier-covered area of more than 400 square miles. Fifteen glaciers in this section have lengths greater than five miles. The longest are Black Rapids and Susitna Glaciers, each 25 miles in length. Almost all of the glaciers in this area are retreating.

Some however, like Black Rapids Glacier, have interrupted their retreats with spectacular surges. Between 1912 and 1936, Black Rapids retreated more than three miles. In September or October 1936, the glacier began to surge. During the winter of 1936-

Trident Glacier descends from 13,832-foot Mount Hayes (center). Mount Deborah, at 12,339 feet, is on the right, while Mount McKinley, in the background, rises 125 miles to the west. (Bob Butterfield)

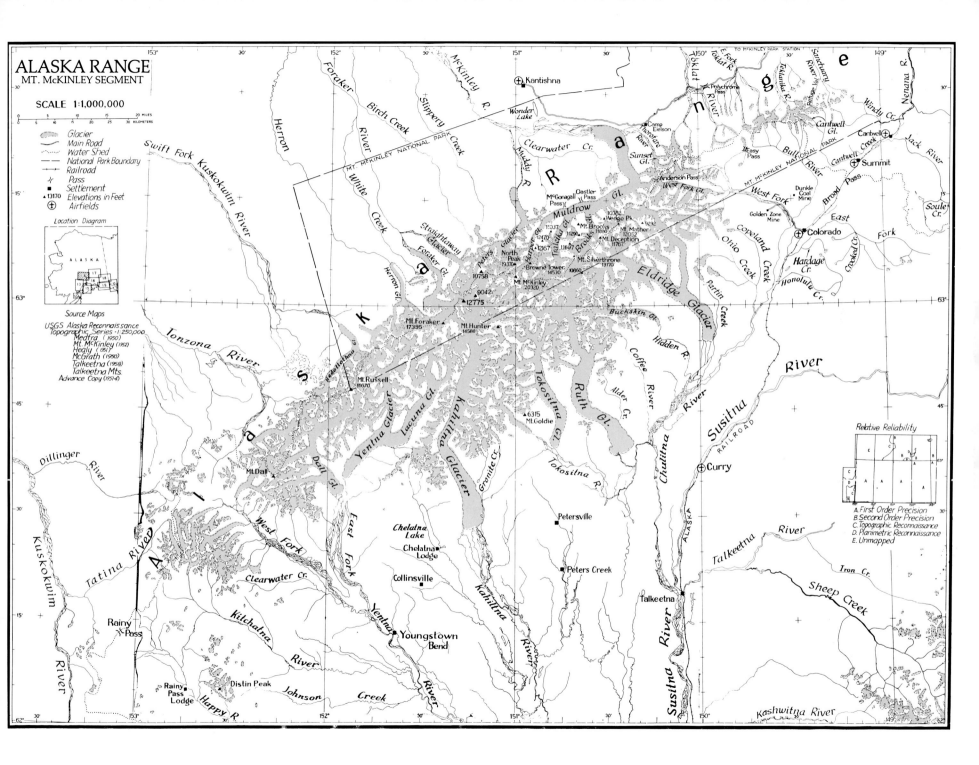

1937 Black Rapids advanced more than three miles in only six months. By February 1937, it had reached a position that threatened the Richardson Highway, adjacent to the Delta River. Fortunately, the glacier advanced no farther. Retreat and downwasting have continued to the present. During the period of maximum surge, Black Rapids Glacier advanced as much as 115 feet per day. As a consequence, it was nicknamed the "Galloping Glacier." A historical marker established at the edge of the Richardson Highway, overlooking Black Rapids Glacier, identifies the terminal moraine from the 1937 surge. The sign itself stands on the moraine of an even earlier and larger surge.

In addition to Black Rapids Glacier, Susitna and Yanert Glaciers have recently surged. In 1952 or 1953, Susitna surged three miles. Yanert, west of Mount Deborah, advanced three miles in 1942 and then surged again in 2000, briefly moving at an estimated 100 yards per day.

South of Black Rapids Glacier, the Clearwater Mountains support a number of small glaciers, most one mile or less in length. The longest, Maclaren Glacier, reaches 10 miles.

The central Alaska Range spans 175 miles from east to west and is almost 50 miles wide. It contains the highest mountain in North America, Mount McKinley (20,320 feet), and Mount Foraker (17,400 feet), the third-highest summit in Alaska. More than 20 glaciers with lengths greater than five miles descend from a broad *névé*. Six glaciers have lengths of 25 miles or more: Kahiltna (43 miles), Muldrow (40 miles), Ruth (36 miles), Yentna-Lacuna (32 miles), Eldridge (30 miles), and Tokositna (25 miles). In 2001 Tokositna has exhibited signs of surging.

Except for Ruth Glacier, which reaches Chulitna Valley, glaciers on the south side of the Alaska Range all terminate in U-shaped valleys that extend up to 30 miles beyond the present ice fronts. On the north side of the range large glaciers extend up to five miles beyond the mountain front and terminate on a low-relief plateau within Denali National Park. All of these glaciers show significant evidence of stagnation, thinning, and retreat.

Muldrow Glacier, with an area of about 200 square miles, flows northeast from high on the slopes of Mount McKinley to a

FACING PAGE: *In the outwash plain of Gulkana Glacier, cottongrass and willows grow in fertile, glacier-produced soil. Gulkana lies 10 miles north of Paxson and east of the Richardson Highway. (Fred Hirschmann)*

RIGHT: *All of the glaciers in the Tordrillo Mountains of the western Alaska Range are retreating, including Capps Glacier, seen here below Mount Spurr, an active volcano. (Fred Hirschmann)*

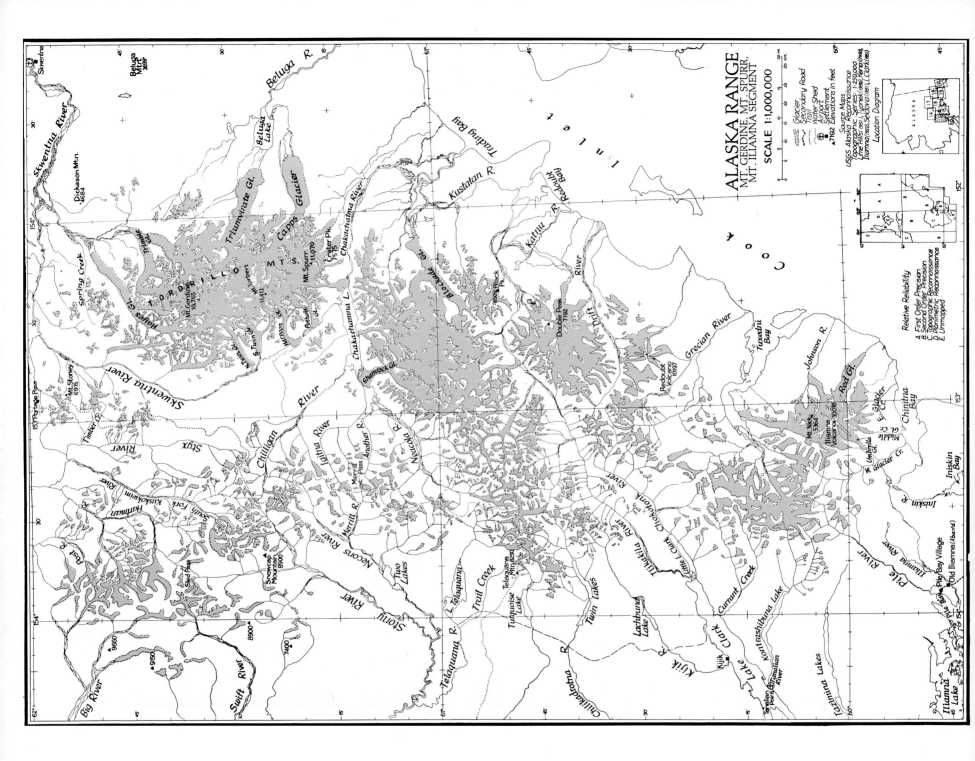

moraine-covered terminus near Eielson Visitor Center. Two named tributary glaciers, Brooks and Traleika, furnish much of Muldrow's ice. Between May 1956 and the summer of 1957, Muldrow Glacier surged four miles. The maximum observed velocity during the peak of movement was about 1,150 feet per day or just under a foot per minute. For almost a decade prior to the surge, a wave of thickening ice moved down the upper part of the glacier at two and one-quarter feet per day. At least four prior surges occurred within the past 200 years, with the most recent pre-1956 surge occurring between 1906 and 1912.

Muldrow Glacier has deposited three sets of terminal moraines during the past 300 years. The largest, visible clearly from Eielson Visitor Center and the Denali park road, was formed after a significant surge, and may represent the maximum ice position of Muldrow Glacier since the start of the Little Ice Age. Ice retreat followed, but about 150 years ago was interrupted by a readvance that deposited a second set of moraines, one to three miles behind the seventeenth-century moraines. The 1957 Muldrow Glacier surge overrode much of the second set of moraines and established a terminal position within three miles of the park road. Today, ablation moraine covers most of the lower five miles of Muldrow Glacier.

The Tordrillo Mountains, which include Mounts Gerdine (11,258 feet) and Spurr (11,070 feet), contain the largest glaciers in the southwestern part of the Alaska Range. Here, Hayes, Trimble, Triumvirate, and Capps Glaciers, all draining eastward, reach lengths of 15 to 25 miles. Eruptions of Mount Spurr in the 1950s and 1990s dusted glaciers with ash as far away as the Chugach Mountains.

All of these glaciers are retreating and stagnating.

West of the Tordrillo Mountains, a number of smaller glaciers lie on the summits and flanks of the Revelation Mountains, and on a number of peaks between Skwentna, Stony and Big Rivers. These are generally valley glaciers with lengths of less than five miles and widths of a mile or less. Revelation Glacier is the largest of the entire group with a length approaching 10 miles. All of the glaciers in this area are retreating and stagnating.

Thermal melting creates a river of mud down glacier-clad Iliamna Volcano, on the west side of Cook Inlet. The volcano's longest glacier is Tuxedni. (Fred Hirschmann)

Aleutian Range

The Aleutian Range, which extends northeast-southwest along the spine of the Alaska Peninsula for more than 500 miles, hosts 27 glaciers with lengths of five miles or more, one exceeding 11 miles. Most glaciers, large and small, are unnamed. Seven areas

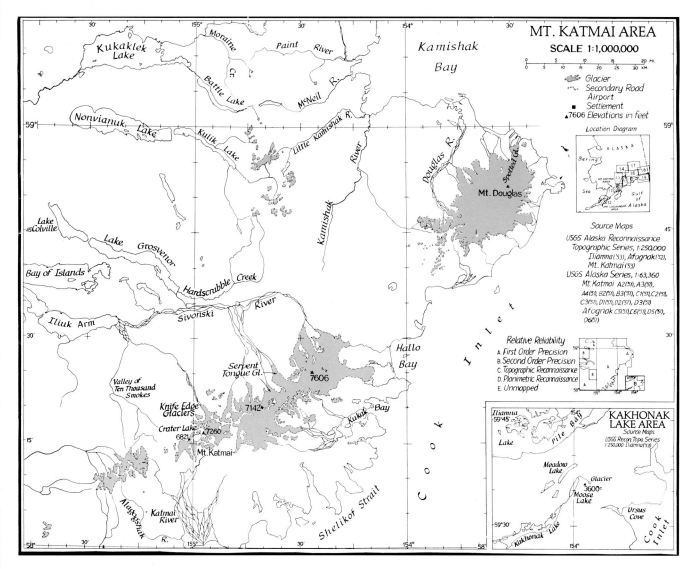

square miles of icefield and valley glaciers. One glacier, Blockade, splits with one terminus damming Blockade Lake and the other flowing into McArthur River, more than 10 miles to the southwest. Another glacier, Shamrock, flows north for 16 miles and pushes its terminus across Chakachamna Lake. Both are rapidly retreating.

More than a dozen unnamed glaciers descend from Redoubt's summit. When the mountain last erupted in 1989 and 1990, much melting of the summit glaciers took place, followed by outburst flash flooding on Drift River. Floods in 1966 covered the lower two miles of Redoubt's northernmost glacier with up to six feet of volcanic debris. Today, the debris-covered terminus is less than 300 feet from Drift River Valley.

Iliamna Volcano has more than a dozen large glaciers that descend from its summit, about four times the area of Redoubt's glaciers. Tuxedni Glacier, the largest with a length of 16 miles, extends from Iliamna's summit to near sea level. Red Glacier, so named for the iron-rich, red-colored moraines that cover its lower three miles, is more than 10 miles long and terminates at an elevation of about 250 feet. Other Iliamna glaciers are Lateral, Johnson, and Umbrella.

Seven glaciers with lengths of five to nine miles occur on the flanks of Mount Douglas (7,064 feet) and Fourpeaked Mountain (6,771 feet), in northeastern Katmai National Park. The largest, Spotted

support glaciers, including the Chigmit Mountains of the northern Aleutian Range; the Kamishak Bay-Big River area; the Mount Katmai area; the Mount Kialagvik-Icy Peak-Mount Chiginagak area southwest of Wide Bay; the Aniakchak Crater area; the Mount Veniaminof area; and the Pavlof Volcano area. Glaciated areas of the Aleutian Range may exceed 1,000 square miles.

The highest summits in the Chigmit Mountains are Redoubt (10,197 feet) and Iliamna (10,016 feet), both glacier-covered, active volcanoes. Additionally, the northern Chigmit Mountains contain more than 200

and Fourpeaked Glaciers, have debris-covered termini and show evidence of recent retreat.

Hundreds of glaciers, 20 of which have lengths of five miles or greater, are located in central and southwestern Katmai National Park. Most descend from a 50-mile-long group of ice-covered volcanoes. Thirteen are more than five miles long, with the longest being 12-mile Hallo Glacier. Prior to its 1912 eruption, Mount Katmai's summit was similar to those of Redoubt and Iliamna Volcanoes, completely encircled by active glaciers. As a result of the eruption, about 2,000 feet of the mountain's summit, including snowfields and glaciers, disappeared, leaving many beheaded glaciers. Since then, two small glaciers have formed in the caldera. Glaciers on Mount Katmai and on adjacent peaks are covered by ash and pumice deposits that exceed tens of feet in thickness in a number of places.

Glaciers on the summits and slopes of Mounts Chiginagak and Kialagvik, and Aniakchak Crater each cover about 10 square miles and are restricted to elevations above 3,000 feet. Only one glacier, four miles in length and unnamed, descends from Mount Kialagvik to near the 1,000-foot level. Its terminus forms the headwaters of Dog Salmon River.

Icy Peak contains over a dozen small unnamed glaciers that descend from an upland, over five miles in length, to below 1,000 feet elevation. One glacier, whose

Bob Gerhard peers into an ice cave melted by a fumarole on the flanks of Mount Redoubt in Lake Clark National Park. Redoubt, at 10,197 feet, is an active volcano that supports several glaciers. (Fred Hirschmann)

meltwater becomes the headwaters of Glacier Creek, stretches more than five miles and terminates at less than 600 feet elevation.

The summit of Mount Veniaminof (7,075 feet) and its southern slopes are covered by a large subcircular glacier larger than 60 square miles. Eight named valley glaciers, Cone, Fog, Island, Outlet, Crab, Harpoon, Finger, and Slim, with a combined area of about 10 square miles, descend from the summit region. A number of other unnamed glaciers are present to the southwest, east, and northeast. The terminus of Slim Glacier extends to an elevation of less than 1,000 feet, more than 11 miles from the summit crater. All of the outlet glaciers show evidence of recent retreat.

The glaciers of the Pavlof Volcano area are the southernmost in the Aleutian Range and cover more than 50 square miles. Much of the upper slopes of Pavlof Volcano (8,250 feet), Little Pavlof Volcano, Double Crater, Pavlof Sister Volcano, Mount Hague, and Mount Emmons are covered by snowfields and ice above 1,000 feet. None of these glaciers have been studied in recent years.

Aleutian Islands

The Aleutian Island chain, which separates the Pacific Ocean from the Bering Sea, contains 46 volcanoes from the Pleistocene or younger, 24 of which have been active since the mid-1700s. It extends more than 1,100 miles from the Alaska Peninsula to Attu Island. At least 10 islands in the eastern and central part of the arc are reported to support glaciers. From east to west, these are Unimak, Akutan, Unalaska, Umnak, Yunaska, Atka, Great Sitkin, Tanaga, Gareloi, and Kiska. For the most part, glaciers of the Aleutian Islands descend from near-summit levels of active and dormant volcanoes, either into their calderas or down their flanks. None reach sea level.

Minimal investigation has been done on Aleutian Islands glaciers during the past 50 years, making current information difficult to find. Research dates have been provided for this section wherever possible.

Unimak Island: Seventy-five mile Unimak Island is the easternmost and largest of the Aleutian Islands. Shishaldin Volcano (9,373 feet), Isanotski Peaks (8,025 feet), and Roundtop Mountain (6,140 feet) on the eastern part of the island host a continuous snow and ice cover nearly 25 miles in length. From 1940s photography and 1972 Landsat images, glaciologists discovered that on the southwestern end of Unimak, Westdahl Volcano (5,118 feet) supports a 10-square-mile glacier ice cap with at least two outlet glaciers. These have

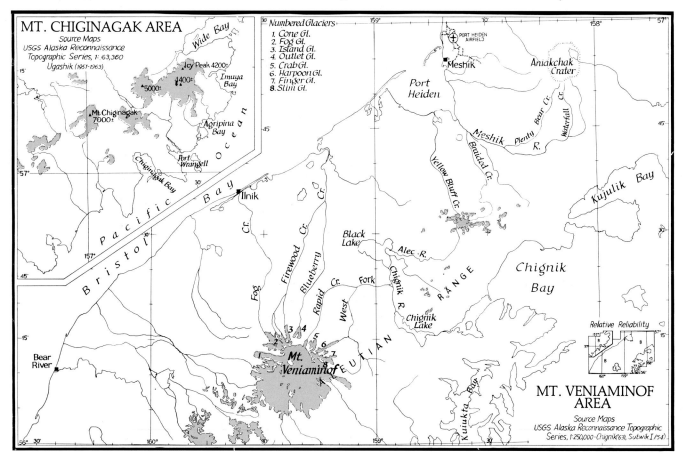

lengths of five-and-one-half miles and seven-and-one-half miles, respectively. An eruption occurred through Westdahl's ice cap in 1992.

Akutan Island: According to 2001 Alaska Volcano Observatory (AVO) reports, a small section of the floor of Akutan's summit caldera is covered with glacier ice.

Unalaska Island: The greatest concentration of glaciers in the Aleutians is on 66-mile-long Unalaska, where 5,905-foot Makushin Volcano is covered by a small ice cap that feeds several small outlet glaciers. In 1996, these descended to about 2,000 feet. Several small valley glaciers occur on the flanks of the Shaler Mountains, located in the southcentral part of the island.

Umnak Island: Field studies from 1959 and Landsat images from 1977 showed that Mount Vsevidof (7,050 feet) supported two valley glaciers while Mount Recheshnoi (6,510 feet) supported at least seven glaciers. At Okmok Caldera, at least one small summit glacier and several hanging glaciers existed inside the south rim. AVO geologists reported that as of 1999, little, if any, ice remained on Umnak's peaks.

Yunaska Island: The summit caldera of Yunaska Island has at least two tongues of ice that descend to the crater floor and at least one hanging glacier. When photographed in 1996, all were less than one-half-mile long.

Atka Island: Two volcanoes on Atka Island, Korovin Volcano (5,029 feet) and Mount Kliuchef (4,760 feet), have summit glaciers, with a glacier-covered area of seven square miles. Field studies from the late 1950s show that one northeast-flowing ice tongue descended from the summit of Mount Kliuchef and terminated at an elevation of about 1,700 feet.

Great Sitkin Island: Much of the west flank of 5,708-foot Great Sitkin Volcano's summit crater is covered by a glacier. Additionally, nine small glaciers of varying lengths descend from its summit. The largest of these glaciers was about two miles long when mapped in the mid-1950s.

Tanaga Island: A 1957 map made by the U.S. Army shows a few small glaciers on 5,925-foot Tanaga Volcano and several smaller glaciers on an unnamed 4,796-foot volcano to the east. The largest glacier on the unnamed volcano heads at 4,280 feet and descends to about 2,900 feet.

Gareloi Island: This island possesses a small glacier-covered area on Mount Gareloi (5,161 feet). At least two small glaciers exist on the north side of its cone. A tiny glacier was also mapped on the southeast side of its summit by USGS in 1954.

Kiska Island: As reported by Keith Henderson and Robert Putnam in a 1947 *Harvard Mountaineering* article, climbers in 1943 described a small, ice-covered area on Kiska Volcano (4,002 feet), as being a "small, decadent, dirty piece of ice, but it still merited the name glacier."

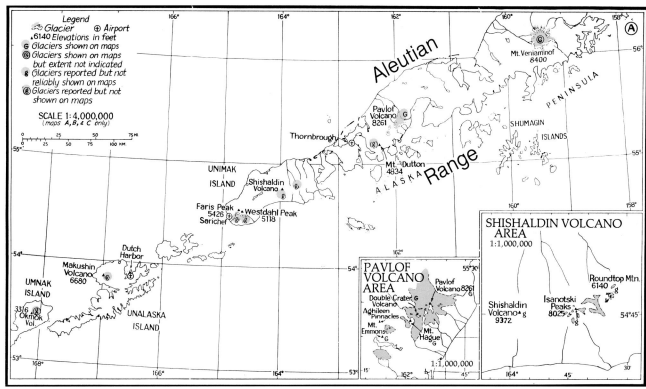

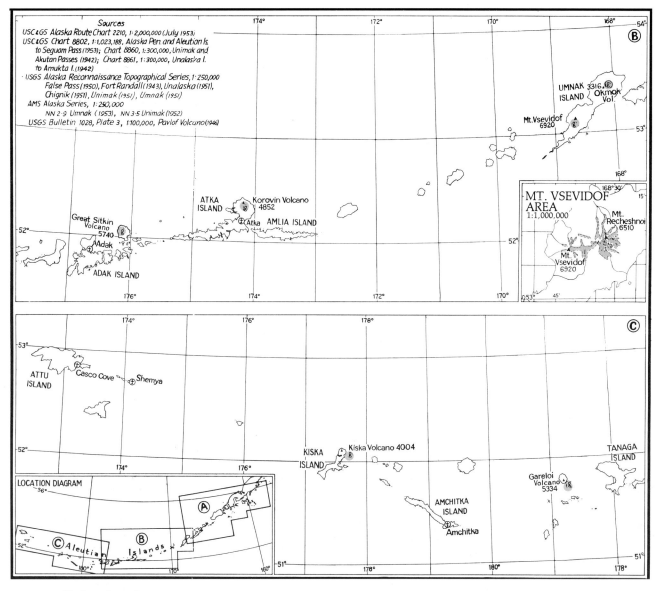

narrow upland region, located in the center of the island between Koniag Peak (4,470 feet) and Mount Glottof (4,405 feet). The largest glacier, Koniag, and the only named one on the island, is a little more than one mile long. All show significant evidence of retreat.

Ahklun and Wood River Mountains

The Ahklun and Wood River Mountains of southwest Alaska are northeast- to southwest-trending mountain ranges, each with a length of about 100 miles and a width of about 30 miles. They extend north from Bristol Bay. Minimally studied, most of what scientists know about them is based on map data collected between 1954 and 1979. Glaciers exist at several locations along the crest of the westcentral part of the mountains. One hundred six were mapped, most small cirque glaciers. At least four larger glaciers, all less than two and one-half miles, are clustered on the north slopes of 5,026-foot Mount Waskey. A fifth glacier, Chikuminuk, the largest in the area, was about three miles long, less than one mile wide, and had an area of less than two square miles when it was mapped during the International Geophysical Year (1957-1958). Then, much of its surface was bare ice with a small area of accumulation. When resurveyed in 1996, Chikuminuk's terminus had retreated more than one-half mile and its area had decreased by about nine percent. However, it was thickening in its accumulation area.

Kodiak Island

Kodiak Island is located in the northcentral Gulf of Alaska, south of Cook Inlet and east of Shelikof Strait. With a length of about 100 miles and a maximum width of nearly 60 miles, Kodiak is the largest island in Alaska. During the Pleistocene, much if not all of Kodiak Island was covered by a large icefield. Numerous fiords, several of which nearly bisect the island, are evidence of the extent of Pleistocene glacier cover. Today 40 cirque glaciers exist in a

Kigluaik Mountains

Today, the only glaciers on the Seward Peninsula are located in the 40-mile east-west-trending Kigluaik Mountains. Three glaciers, with an area of less than a square mile, exist in drainages on the flanks of 4,714-foot Mount Osborn and adjacent peaks, 30 miles north of Nome. During the twentieth century, several other glaciers completely disappeared. The largest existing glacier, Grand Union in the Grand Union Creek drainage, is less than one-half-mile

Ice melts in the crater of Mount Veniaminof as steam escapes from the volcano's vent. Alaska Volcano Observatory monitors volcanic activity around the state, but also gathers information on associated glaciers. (Courtesy Tina Neal, Alaska Volcano Observatory)

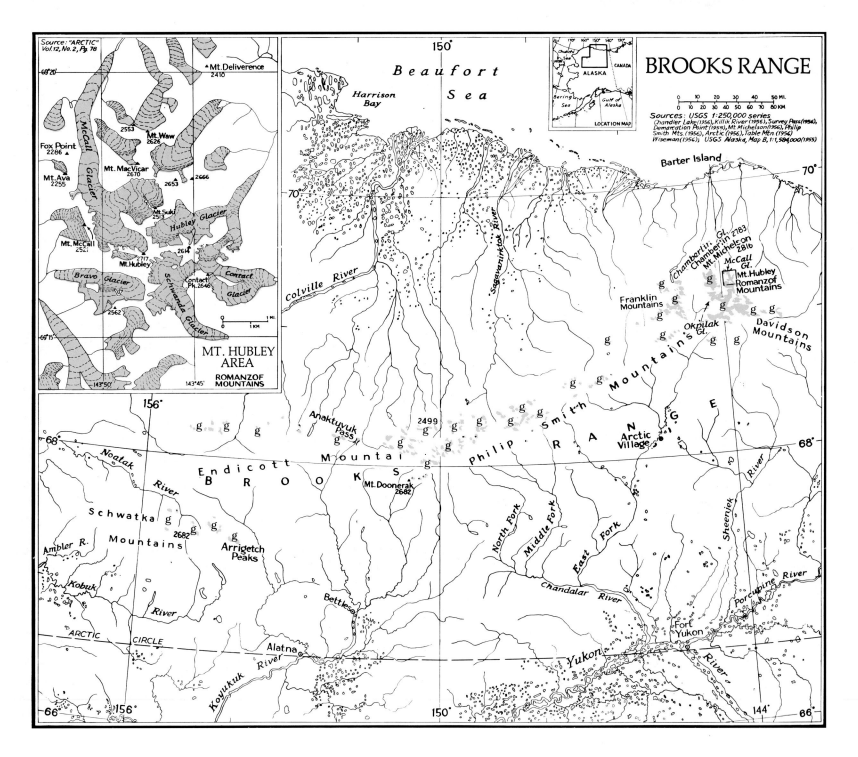

long. Smaller, stagnating, cirque glaciers, informally called Thrush and Phalarope, occur in north-facing, canyonlike cirques. In 1986, Thrush was about 1,000 feet long. Phalarope, located about five miles northwest of Thrush, is the smallest of the three, with a 1986 length of about 500 feet. At the rate these glaciers were retreating and thinning, it is possible that one or more of them may soon disappear.

Brooks Range

The Brooks Range, the northernmost mountain range in Alaska, extends 600 miles in an east-west direction from Canada to the Chukchi Sea. The Brooks Range consists of 11 mountain groups: British, Davidson, Romanzof, Franklin, Sadlerochit, Shublick, Philip Smith, Endicott, Schwatka, Baird, and DeLong. These ranges form the drainage divide between the Arctic Slope to the north and Kobuk and Yukon Rivers to the south. The highest mountains in the Brooks Range are the Romanzofs, with Mounts Chamberlin (9,020 feet), Isto (9,050 feet), and Michelson (8,855 feet). Glaciers exist within the Romanzof, Franklin, Philip Smith, Endicott, and Schwatka Mountains, with the largest glaciers and the greatest concentration of glacier ice in the east. As the entire Brooks Range is located north of the Arctic Circle, glaciers are generally subpolar. Most are cirque glaciers with a northern exposure.

Little was known about Brooks Range glaciers until after World War II. A 1930 report stated that only three glaciers had been seen in all of northwest Alaska. In July 1911, Philip Smith made photographs of two of these glaciers in the Arrigetch Peaks area. In the eastern Brooks Range, the only description of glaciers was limited to observations of Ernest Leffingwell in June 1907.

The Romanzof Mountains have a glacier-covered area of about 100 square miles, and at least five glaciers with lengths of five miles or more. Two, Okpilak and McCall, have been investigated for many years. Okpilak, visited by Leffingwell in 1907, is about five miles long. Between 1907 and 1958, it retreated nearly one-quarter of a mile, a trend that continued through the end of the century. All glaciers in this region that reach below 5,500 feet show evidence of retreat and thinning.

McCall Glacier, 10 miles northeast of Okpilak and the second-largest in the Brooks Range, heads on 8,915-foot Mount Hubley. Because of its size and relatively easy access, McCall is the most intensively studied glacier in the range. Scientists have analyzed its heat flow, ice, water, and mass balance. During a four-year period of the International Hydrological Decade (1969-1972), McCall's mass balance was consistently negative with 98 percent of ice loss due to melting.

Exhibiting the silty, green cast of glacially fed lakes, Lake Chikuminuk lies in Wood-Tikchik State Park, north of Dillingham. Small cirque glaciers dot the mountains here. (Hugh S. Rose)

LEFT: *With a Brooks Range glacier as backdrop, a hiker ascends a ridge between the Jago and Aichilik Rivers in the Romanzof Mountains of the Arctic National Wildlife Refuge. (Brian Okonek)*

FACING PAGE: *Sharp* arêtes *atop granite walls are reminders of glacial scouring lasting thousands of years in the Arrigetch Peaks in Gates of the Arctic National Park. Brooks Range glaciers today are in retreat. (Hugh S. Rose)*

Mount Isto, Tugak Peak (8,105 feet), and Mount Michelson, also in the Romanzofs, support large valley glaciers. The summit of Mount Chamberlin, in the Franklin Mountains, is also glacier-covered; the two largest glaciers, Chamberlin on the west and an unnamed glacier on the east, extend less than two miles.

Near Atigun Pass, 133 glaciers have been identified, the largest being about one-and-one-half-miles long. All of the glaciers examined had their termini at elevations of 5,000 feet or greater. All were wasting away. During a two-month summer period of observation, the entire previous year's snow accumulation, as well as an additional three to six feet of ice, was lost at three cirque glaciers near Atigun Pass.

In the Anaktuvuk Pass area, many of the glaciers lie in deep, north-facing cirques at elevations below the present regional snowline. Here, they show evidence of twentieth-century thinning and retreat and are in a state of disequilibrium with present climatic conditions.

In 1962, Philip Smith's 1911 Endicott Mountains photographic stations were reoccupied and photographic comparisons of glacier position and glacier health in the Arrigetch Peaks region were made. In the 51 years between studies a recession and thinning of the glaciers in the region was noted. Today, most glaciers in the area are two miles or less in length and restricted to elevations of 6,000 feet or greater. All show evidence of twentieth-century thinning and retreat.

In the Schwatka Mountains, south of the Noatak River, at least six peaks support valley glaciers. The largest glaciers descend from the slopes of 8,510-foot Mount Igikpak and 7,310-foot Oyukak Mountain. The largest glacier on Mount Igikpak is two miles long, unnamed, and terminates at an elevation of about 3,800 feet.

Where to See Alaska Glaciers

By Bruce Molnia

With as many as 100,000 glaciers in Alaska, one might think seeing them is easy. For a few glaciers, this is true. Some, such as Mendenhall, are accessible by public transportation. Others, such as those of the eastern Aleutian Islands, are so remote it requires months of preparation, complicated logistics, favorable weather, and great expense to see them. The number of glaciers a person can see in Alaska is a matter of time, cost, and effort; access is by road, foot, air, and water.

Although most of the state is not accessible by road, many glaciers are. Worthington Glacier (Richardson Highway), Matanuska Glacier (Glenn Highway), Portage and Exit Glaciers (Seward Highway), Childs and Sheridan Glaciers (Copper River Road), and Mendenhall Glacier (near Juneau) are close to frequently traveled roads and are safely visited by car. Charter buses make daily trips to Portage, Exit, and Mendenhall Glaciers. All offer opportunities for hiking and a close look at a glacier in action.

Those enthusiasts with strong legs and an outdoor bent can hike to Muldrow Glacier in Denali National Park, Valdez Glacier, many Chugach Mountains glaciers north and east of Anchorage, several Kenai Mountains glaciers south of Anchorage, and glaciers along the Copper River. U.S. Forest Service trails lead to many of the Juneau Icefield's outlet glaciers.

Glaciers can also be seen from windows of commercial airliners. Many routes overfly some of the most spectacular glaciers in Alaska. The Seattle to Anchorage route flies directly over Bering Glacier and then crosses Prince William Sound. The Seattle to Juneau route traverses nearly the entire length of the Coast Mountains. Flights on the Chicago to Anchorage and Detroit to Anchorage route pass over parts of the St. Elias and Chugach Mountains. The Anchorage to Fairbanks route crosses Denali National Park and many of its large glaciers. Perhaps

Kayaking to and camping near glaciers affords those who love the outdoors a respite from crowded summer campgrounds. These children relax near Lamplugh Glacier, in Glacier Bay National Park. (Roy M. Corral)

the most remarkable commercial route is the Anchorage to Juneau milk run, which stops in Cordova and Yakutat. It passes over Prince William Sound, Bering Glacier, Icy Bay, Malaspina Glacier, Yakutat Bay, Glacier Bay, and part of the Coast Mountains. Between Juneau and Yakutat, spectacular views of large valley glaciers and glacier-covered summits like Mounts Fairweather and Lituya are available on both sides of the aircraft.

Charter air services offer scenic flights over Glacier Bay; the Sargent, Harding, Juneau, and Stikine Icefields; the Chugach and St. Elias Mountains; and the Alaska Range. Some flying services land ski-equipped, single-engine aircraft on a number of glaciers. One Talkeetna-based operator offers four different flightseeing tours of Mount McKinley and Denali National Park.

Another way to get close to Alaska's glaciers is by helicopter. Several operators present flightseeing tours and glacier landings. In Juneau, as many as 50 helicopter landings a day occur on Mendenhall Glacier.

Cruise ships depart Seattle, Vancouver, B.C., and many Southeast Alaska cities daily from late spring to early fall and sail to destinations including Glacier Bay, Yakutat Bay, and Prince William Sound. Ferries of the Alaska Marine Highway system sail past glaciers of the Inside Passage and visit several in Prince William Sound.

Concessionaires and guides also provide opportunities to sail to tidewater glaciers in Prince William Sound, Glacier Bay, and Frederick Sound. Day cruises from Whittier, Seward, Juneau, Ketchikan, and Petersburg visit nearby fiords where passengers watch glaciers calve icebergs. One tour boat operator advertised that 100 glaciers could be seen in a single day. In some locations, the number is significantly higher.

To become truly intimate with an Alaska glacier requires an individual take the time to sit and watch or to walk nearby and contemplate the glacier's magnificence. There are many places to commune with glaciers and their surroundings. U.S. Forest Service visitor centers at Mendenhall and Portage Glaciers provide diverse information and ranger-guided hikes. Large National Park Service visitor centers at Kenai Fiords, Wrangell-St. Elias, and Denali National Parks, and smaller U.S. Forest Service and State Park system information centers at Childs, Exit, and Worthington Glaciers provide information about local glaciers and their history. Every year, new facilities are constructed and improved informational displays are prepared.

The opportunities are endless. Go out and meet a glacier.

Mount Hunter, in the Alaska Range, towers over climbers ferrying gear to Camp 2 from the base camp and airstrip on the southeast fork of Kahiltna Glacier. Flightseeing tours around Mount McKinley operate daily, depending on the weather, from near Denali Park Headquarters or Talkeetna. (John Schwieder)

Glossary

Ablation: loss of ice and snow from a glacier system. This occurs through a variety of processes including melting and runoff, sublimation, evaporation, calving, and wind transportation of snow out of a glacier basin.

Arêtes, Alaska Range. (Bob Butterfield)

Accumulation: the addition of ice and snow to a glacier system. This occurs through a variety of processes including precipitation, firnification, and wind transportation of snow into a glacier basin from an adjacent area.

Altithermal: the part of the Holocene between 5,300 and 6,600 years ago. During this interval, climate was drier and warmer than at present, with temperatures three to five degrees Fahrenheit warmer than today. Glaciers throughout Alaska retreated during the Altithermal.

Arête: a jagged, narrow ridge that separates two adjacent glacier valleys or cirques. The ridge frequently resembles the blade of a serrated knife. The term is French and refers to the bones in a fish backbone.

Bergschrund: a single, large crevasse or a series of subparallel crevasses that develop at the head of a glacier. Generally, this is where moving ice pulls away from the bedrock wall of the cirque against which it accumulated. During the winter, the crevasse may fill with snow. In spring or summer, it reopens and becomes visible. The term is German and means mountain crevice.

Bergy seltzer (ice sizzle): a crackling or sizzling similar to that made by champagne, seltzer water, or Rice Krispies cereal, but louder. The sound is made as air bubbles that formed at different pressures are released during the melting of glacier ice.

Calving: the process by which pieces of glacier ice break away from the terminus of a glacier that ends in a body of water. Once they enter the water, the pieces are called icebergs.

Cirque: a bowl-shaped, amphitheater-like depression eroded into the head or the side of a glacier valley. Typically, a cirque has a lip at its lower end. The term is French and is derived from the Latin word *circus*, meaning circle.

Bergschrund. (Bob Butterfield)

Cirque glacier: a small glacier that forms within a cirque basin, generally high on the side of a mountain.

Crevasse: a crack or series of cracks that develops in the surface of a moving glacier in response to differential stresses caused by glacier flow. Crevasses may range in shape from linear to arcuate, in length from several feet to more than a mile, and may have any orientation with respect to the flow direction of a glacier. The deepest crevasses may exceed 100 feet.

Crevasses, Bering Glacier. (Bruce Molnia)

Distributary: a tongue of glacier ice that flows away from the main trunk of the glacier. This results from differential melting changing the gradient of part of a glacier.

Downwasting: the thinning of a glacier due to the ablation of ice. This loss of thickness may occur in both moving and stagnant ice.

Downwasting, Triumvirate Glacier. (Fred Hirschmann)

Drift: a term used to describe all types of glacial sedimentary deposits, regardless of the size or degree of sorting of the material involved. The term includes all sediment that is transported by a glacier, deposited directly by a glacier, or deposited indirectly by running water that originates from a glacier.

Drift, Muldrow Glacier. (Hugh S. Rose)

Erratic: a rock of unspecified shape and size transported a significant distance from its origin by a glacier or iceberg and deposited by melting of the ice. Erratics range from pebble-size to larger than a house and are usually of a different composition than the bedrock on which they are deposited.

Eskers near Casement Glacier, Glacier Bay National Park. (Bruce Molnia)

Esker: a meandering, water-deposited, steep-sided, sediment ridge that forms within a sub-glacial or englacial stream channel. Its floor can be bedrock, sediment, or ice. Subsequent melting of the glacier exposes the deposit. Composed of stratified sand and gravel, eskers can range from feet to miles in length and may exceed 100 feet in height.

Eustacy: the worldwide sea level regime and its fluctuations caused by changes in the quantity of seawater available, generally caused by water being added to or removed from glaciers.

Fiord: a glacially eroded or modified U-shaped valley that extends below sea level into a deep bay or onto the continental shelf. Depths frequently reach more than 1,000 feet below sea level; widths can extend up to five miles, lengths to more than 100 miles. Tidewater glaciers commonly calve icebergs into a fiord's waterway.

Firn: an intermediate stage in the transformation of snow to glacier ice. Snow becomes firn when it has been compressed so that no pore space remains, a process that takes less than a year.

ALASKA GEOGRAPHIC® 103

Firn line: a line from glacier edge to glacier edge that marks the transition between exposed glacier ice and the snow-covered surface of a glacier. During the summer melt season, this line moves up-glacier. At the end of the melt season the firn line separates the accumulation zone from the ablation zone.

Firn line, Chugach Mountains. (Bruce Molnia)

Foliation: the layering or banding that develops in a glacier during the process of transformation of snow to glacier ice. Individual layers, called folia, are visible because of differences in crystal or grain size or because of an alternation of clear ice and bubbly ice.

Glacial stream: a channel of flowing liquid water on, in, or under (supraglacial, englacial, or subglacial respectively) a glacier, moving under the influence of gravity.

Glacier: a large, perennial accumulation of ice, snow, rock, sediment, and liquid water originating on land and moving down-slope under the influence of its own weight and gravity; a dynamic river of ice.

Foliation, Lamplugh Glacier. (Bruce Molnia)

Glacial stream, Shoup Glacier. (Nick Jans)

Hanging glacier: a glacier that originates high on the wall of a glacier valley and descends only part of the way to the surface of the main glacier. Avalanches and icefalls are the mechanisms for transferring ice and snow to the valley floor below.

Hanging glacier, Cathedral Spires. (Fred Hirschmann)

Holocene: the current epoch of geologic time. The Holocene began about 12,000 years ago, at the end of the Pleistocene epoch.

Horn, Michaels Sword. (John Hyde)

Horn: a rocky, pointed, mountain peak, generally pyramidal in shape, bounded by the walls of three or more cirques. Headward erosion has cut prominent faces and ridges into the peak. When a peak has four symmetrical faces, it is called a matterhorn, after the famous peak in the European Alps.

Iceberg: a block of ice that has broken or calved from the face of a glacier and is floating in a body of marine or fresh water. Alaska icebergs rarely exceed 500 feet in maximum dimension.

Iceberg, Chenega Glacier. (Hugh S. Rose)

Ice cap: a dome-shaped or plate-like cover of perennial snow and glacier ice that completely covers the summits of a mountain mass so that no peaks emerge through it. The term also applies to a continuous cover of snow and ice on an Arctic or Antarctic land mass that spreads outward in all directions because of its mass. Ice caps have areas of less than about 20,000 square miles.

Icefield: a mountainous area where large interconnecting valley glaciers are separated by mountain peaks and ridges that project through the ice as nunataks. The lower parts of the valley glaciers serve as outlet glaciers and drain ice from the icefield. Alaska icefields include the Stikine, Juneau, Harding, and Sargent; each has an area of more than 500 square miles.

Icefall, Lowell Glacier. (Michael R. Speaks)

Icefall: part of a glacier where ice flows over a bed with a very steep gradient. As a result, the surface of the glacier is fractured and heavily crevassed. In a river system, this would be a waterfall.

Isostacy: changes within Earth's surface where material within the crust and mantle is displaced in response to the increase (isostatic depression) or decrease (isostatic rebound) in mass at any point on Earth. This change is often caused by advance or retreat of glaciers.

Jokulhlaup: a glacier outburst flood resulting from the failure of a glacier-ice-dam, glacier-sediment-dam, or from the melting of glacier ice by a volcanic eruption. The term is Icelandic.

Kame: a sand and gravel body, ranging from poorly sorted to well stratified, formed by running water depositing sediment on stagnant or moving glacier ice. Kames form within crevasses as crevasse fills or crevasse ridges, on flat or inclined glacier surfaces, in holes, or in cracks in glacier ice. When a kame forms between a glacier and adjacent land surface, a kame terrace results. Kame shapes include hills, mounds, knobs, hummocks, or ridges.

Kettle: a depression that forms in an outwash plain or other glacial deposit by the melting of an in-situ block of glacier ice that was separated from the retreating

Kame, Herbert Glacier. (Bruce Molnia)

Jokulhlaup. (Bruce Molnia)

glacier margin and subsequently buried by glacier sediment. As the buried ice melts, the depression enlarges. A pit pond is a similar-looking feature, formed by the melting of a block of ice floated to its depositional site by meltwater and subsequently buried by sediment. As it melts, a depression in the surface of the outwash plain develops.

Kettles. (Bruce Molnia)

ALASKA GEOGRAPHIC® 105

Little Ice Age (Neoglaciation): the most recent period of temperate glacier expansion and advance. Characterized by a cooling of three or more degrees Fahrenheit, the Little Ice Age began approximately 650 years ago and continued into the twentieth century in many locations. Temperate glaciers in North America, South America, Africa, Europe, and Asia were affected.

Loess: a deposit of fine-grained, windblown, glacial dust.

Mass balance: sum of the accumulation and ablation on a glacier during one year.

Lateral moraine, Barnard Glacier. (AGS file photo)

Moraine: a general term for unstratified and unsorted deposits of sediment that form through the direct action of, or contact with, glacier ice. Many different varieties are recognized based on their position with respect to a glacier.

• *ablation moraine (ablation till)*: an irregular-shaped layer or pile of glacier sediment formed by the melting of a block of stagnant ice. Ablation moraines form on former glacier beds.

• *ground moraine*: a blanket of glacier till deposited on all of the surfaces over which a glacier moved.

• *lateral moraine*: a generally linear accumulation of glacial sediment, located on the ice surface at the edge of a glacier. It extends in an up-glacier to down-glacier direction. In a moving glacier, it is a continuous, linear sediment ridge located at the junction of the ice margin and the valley wall. It forms by the accumulation of rock material falling onto the glacier from the valley wall, rather than by water deposition.

• *medial moraine*: a linear accumulation of glacial sediment on the center of the ice surface, parallel to the glacier's flow. A large valley glacier may have more than a dozen medial moraines,

Medial moraines, Barnard Glacier. (AGS file photo)

each formed by the joining of two lateral moraines during the merging of two tributary glaciers.

• *push moraine*: a pile of unstratified glacial sediment that is formed in front of the ice margin by the terminus of a glacier bulldozing sediment in its path as it advances.

Push moraine, Harriman Glacier. (Bruce Molnia)

• *recessional moraine*: an accumulation of glacial sediment that forms when the terminus of a retreating glacier remains at or near a single location for a period of time sufficient for a cross-valley ridgelike deposit to form. A series of such moraines represents a number of periods of relative stability during a glacier's retreat.

Recessional moraine, Bering Glacier. (Bruce Molnia)

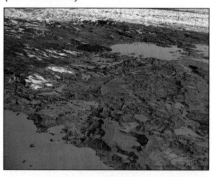

106 ALASKA GEOGRAPHIC®

• *terminal moraine (end moraine)*: a cross-valley, ridgelike accumulation of glacial sediment that forms at the farthest point reached by the terminus of an advancing glacier.

Ogive: an arcuate, convex band or undulation that forms on the surface of a glacier at the base of an icefall. Two types of ogives occur: wave ogives, undulations of varying height; and band ogives, alternating light- and dark-colored bands.

Outwash plain: a broad, low-slope, alluvial plain composed of glacially eroded, sorted sediment that has been transported by meltwater. The alluvial plain begins at the foot of a glacier and may extend for several miles. Typically, its sediment becomes finer-grained with increasing distance from the glacier terminus.

Piedmont glacier: a broad, fan- or lobe-shaped glacier, located at the front of a mountain range. It forms when one or more valley glaciers flow from a confined valley onto a plain where they can expand. The largest piedmont glacier in Alaska, Malaspina, is more than 30 miles wide.

Pleistocene: the epoch of geologic time that began approximately 2.5 million years ago and ended about 12,000 years ago. During this interval continental glaciers formed and covered significant parts of Earth's surface. This epoch is informally called The Great Ice Age or the Glacial Epoch. Together, the Holocene and Pleistocene epochs comprise the Quaternary Period.

Plucking: the mechanical removal of pieces of rock from a bedrock face that is in contact with glacier ice. Blocks are quarried and prepared for removal by the freezing and thawing of water in cracks, joints, and fractures. The resulting pieces are frozen into the glacier ice and transported away.

Reconstituted glacier (*glacier remanié*): a glacier formed below the terminus of a hanging glacier by the accumulation and reconstitution by pressure melting (regelation), of ice blocks that have fallen and/or avalanched from the terminus of the hanging glacier.

Regelation: the process through which water, derived from ice that melted under pressure, is refrozen, when the pressure is relieved.

***Roche moutonnée*:** an elongated, rounded, asymmetrical, bedrock knob produced by glacial erosion. It has a gentle slope on its up-glacier side and a steep to vertical face on the down-glacier side.

Moulin, La Perouse Glacier. (Bruce Molnia)

Outwash plain, Casement Glacier. (Bruce Molnia)

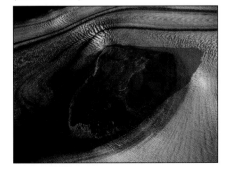

Nunatak, Grand Plateau Glacier. (R.E. Johnson)

***Moulin* (glacier mill):** a narrow, tubular chute or crevasse through which water enters a glacier from the surface. Occasionally, the lower end of a *moulin* may be exposed in the face of a glacier or at the edge of a stagnant block of ice.

***Névé* (accumulation zone):** the area of a glacier covered with perennial snow.

Nunatak: a mountain peak or ridge that pokes through the surface of a glacier, separating adjacent valley glaciers. The term is Greenlandic.

Rock flour (glacial flour): fine-grained, silt-sized sediment formed by the mechanical erosion of bedrock at the base and sides of a glacier by moving ice. When it enters a stream, it turns the stream's color brown, gray, iridescent blue-green, or milky white (glacier milk).

Rock glacier: a glacierlike landform that frequently heads in a cirque and consists of a valley-filling accumulation of angular rock blocks. A rock glacier resembles a glacier in shape, but has little or no visible ice at the surface. Ice may fill the spaces between rock blocks. Some rock glaciers move, although very slowly.

Rock flour. (Bruce Molnia)

Rock glaciers. (Bob Butterfield)

Séracs, Excelsior Glacier. (Curvin Metzler)

Sérac: a jagged pinnacle or tower of glacier ice, located at the surface of a glacier. Frequently the surface of a large area of a glacier will be covered by seracs.

Suncups: a series of bowl-like depressions melted into a snow surface, separated by a network of connected ridges. Individual suncups may be more than three feet deep and ten feet in diameter. Suncups form during warm, sunny conditions.

Surge: a short-lived, frequently large-scale increase in the rate of movement of the ice within a glacier. Ice velocities may increase 10 to 100 times above normal flow rates. Some surges cause the terminus of a glacier to rapidly advance. Although not all glaciers surge, those that do often experience surge events with some sort of periodicity.

Tarn: a lake that develops in the basin of a cirque, generally after the disappearance of the glacier.

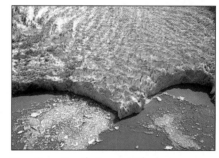
Terminus, South Sawyer Glacier. (Pieter Folkens)

Terminus (toe, snout): the lowermost margin, end, or extremity of a glacier.

Tidewater glacier: a glacier whose terminus ends in saltwater at the head of a fiord, or ends in the ocean.

Till: any unsorted and unstratified accumulation of glacial sediment, deposited directly by a glacier, that has not been reworked by water.

Surge, Malaspina Glacier. (R.E. Johnson)

Till is a heterogenous mixture of different-sized material deposited by moving ice (lodgement till) or by melting in place of stagnant ice (ablation till).

Trimline: a clear boundary line on the wall or walls of a glacier valley that indicates the maximum recent thickness of a glacier. The line may result from a change in the color of the bedrock, indicating the separation of weathered from unweathered bedrock; may be the limit of a former lateral moraine or other sediment deposit; or may be the boundary between vegetation-covered bedrock and bare bedrock.

U-shaped valley: a valley with a parabolic or "U" shaped cross-section, steep walls and a broad, flat floor. Formed by glacial erosion, a U-shaped valley generally results when a glacier widens and oversteepens a V-shaped stream valley.

Valley glacier: a glacier which flows in a bedrock-confined valley.

Bibliography

Bailey, Ronald H. *Glacier*. Planet Earth series. Alexandria, Va.: Time-Life Books, 1982.

Bauer, Peggy, and Erwin Bauer. *Glacier Bay: The Wild Beauty of Glacier Bay National Park*. Seattle: Sasquatch Books, 2000.

Benn, Douglas I., and David J.A. Evans. *Glaciers & Glaciation*. London: Edward Arnold, 1998.

Burroughs, John, John Muir, and others. *Alaska: The Harriman Expedition, 1899*. Two volumes bound as one. New York: Dover, 1986.

Connor, Cathy, and Daniel O'Haire. *Roadside Geology of Alaska*. Missoula: Mountain Press, 1988.

Corral, Kimberly. *A Child's Glacier Bay*. Photos by Roy Corral. Anchorage: Alaska Northwest Books, 1998.

Dufresne, Jim. *Glacier Bay National Park: A Backcountry Guide to the Glaciers and Beyond*. Seattle: Mountaineers Books, 1987.

Ferguson, Sue A. *Glaciers of North America: A Field Guide*. Golden, Colo.: Fulcrum, 1992.

Fowler, Brenda. *Iceman: Uncovering the Life and Times of a Prehistoric Man Found in an Alpine Glacier*. New York: Random House, 2000.

Gallant, Roy A. *Glaciers*. Earth and Sky Science. First Books series. New York: Franklin Watts, 1999.

Howard, Jim. *Guide to Sea Kayaking in Southeast Alaska: The Best Trips and Tours from Misty Fjords to Glacier Bay*. Regional Sea Kayaking Series. Old Saybrook, Conn.: Globe Pequot Press, 1999.

Jettmar, Karen. *Alaska's Glacier Bay: a traveler's guide*. Anchorage: Alaska Northwest Books, 1997.

Kelley, Mark. *Glacier Bay National Park, Alaska*. Text by Sherry Simpson. Juneau: Mark Kelley Photography, 2000.

Knight, Peter. *Glaciers*. Cheltenham: Stanley Thornes, 1999.

Krimmel, Robert M., and Mark F. Meier, eds. *Glaciers and Glaciology of Alaska: Anchorage to Juneau, Alaska, July 21-29, 1989*. International Geological Congress Field Trip Guidebooks Series. Washington, D.C.: American Geophysical Union, 1989.

Lethcoe, Nancy R. *An Observer's Guide to the Glaciers of Prince William Sound, Alaska*. Preface by William O. Field. Illustrated by R. James Lethcoe. Valdez: Prince William Sound Books, 1987.

Miller, Lance D., Joan W. Miller, and Maynard M. Miller, Ph.D. *The Juneau Icefield — Alaska: Color Slides and Text*. Juneau: The Foundation for Glacier and Environmental Research, 1987.

Molnia, Bruce. *Alaska's Glaciers*. Vol. 9, No. 1. Rev. ed. Edited by Penny Rennick. Anchorage: Alaska Geographic Society, 1993.

Post, Austin, and Edward R. LaChapelle. *Glacier Ice*. Rev. ed. Seattle: University of Washington Press in association with The International Glaciological Society, Cambridge, 2000.

Selters, Andrew. *Glacier Travel & Crevasse Rescue: Reading Glaciers, Team Travel, Crevasse Rescue Techniques, Routefinding, Expedition Skills*. 2nd ed. Seattle: Mountaineers Books, 1999.

Sharp, Robert P. *Living Ice: Understanding Glaciers and Glaciation*. Cambridge: Cambridge University Press, 1988.

Skillman, Don. *Adventure Kayaking: Trips in Glacier Bay*. Berkeley: Wilderness Press, 1998.

Williams, Richard S. Jr., and Jane G. Ferrigno. *1989-2001, Satellite Image Atlas of Glaciers of the World: Chapters A-K*. Washington, D.C.: USGPO, n.d.

www.asf.alaska.edu:2222/ (Alaska Synthetic Aperature Radar Facility — ASF, "Glacier Power")

www.avo.alaska.edu/ (Alaska Volcano Observatory — AVO)

www.crevassezone.org/ (The Crevasse Zone)

www.dnr.state.ak.us/mine_wat/water/glacier.htm (State of Alaska, Dept. of Natural Resources, Division of Mining, Land and Water, "Glacier Ice Harvesting in Alaska")

www.gi.alaska.edu/snowice/ (Geophysical Institute, University of Alaska, Snow, Ice, and Permafrost Group)

Index

Ablation 16, 23, 38, 102-104, 106
Accumulation 14, 15, 38, 39, 62, 82, 94, 98, 102, 104, 106, 107
Advance 20, 29, 30, 32, 38, 45, 60, 62, 67, 69, 70, 71, 75, 77, 78, 80, 84, 87, 89, 105, 107, 108
Ahklun Mountains 8, 9, 94
Aialik Bay 10, 81
Alaska Department of Natural Resources (DNR) 2, 10
Alaska Peninsula 8, 82, 89, 92
Alaska Quaternary Center 34, 44
Alaska Range 8, 9, 15, 19, 30, 34, 35, 52, 82-85, 87-89, 101, 102
Alaska Volcano Observatory (AVO) 2, 93, 95
Aleutian Islands 2, 8, 9, 49, 92-94, 100
Aleutian Range 9, 39, 89, 90, 92, 93
Alexander Archipelago 8, 9, 32, 59
Altithermal 32, 102
Amedeo Di Savoia, Prince Luigi (Duke of Abruzzi) 49
American Geographical Society 50, 53, 64
Anchorage 5, 6, 11, 21, 31, 47, 56, 77, 78, 81, 100, 101
Andrews, C.L. 49
Arctic Institute of North America 53
Arête 21, 23, 98, 102
Arrigetch Peaks 97, 98
Artifacts 42, 43

Bagley Icevalley 73
Barnard, Edward Chester 2

Barry Arm 76, 77
Beeman, Susan 10-11
Behnert, Rai 57
Belcher, Sir Edward 70
Benedek, Kristin 42
Bergy seltzer 10, 15, 16, 102
Bergschrund 29, 102
Bering Land Bridge 34, 35
Bering, Vitus 48
Beschel, Roland 55
Blackstone Bay 51, 76, 80
Blake, William 48
Bloom, Art 57
Bohn, Dave 50, 51
Brooks, Alfred Hulse 49, 82
Brooks Range 8, 9, 30, 32, 34, 35, 96-98
Burroughs, John 47

Calving 5, 7, 10, 15, 17, 20, 22, 27, 29, 31, 38, 47, 50, 54, 61, 62, 67-71, 75-77, 101-104
Capps, Stephen 82
Chigmit Mountains 32, 90
Christy, Scott 2
Chugach Mountains 8, 9, 13, 20, 27, 30, 38, 50, 52, 73, 76, 78, 79, 82, 89, 100, 101, 104
Cirque 15, 21, 23, 27-29, 97, 98, 102, 104, 108
Climate 31, 32, 34, 44, 45, 53, 64, 98, 102
Coast Mountains 8, 9, 16, 30, 43, 59, 60-63, 100, 101
College Fiord 29, 47, 50, 71, 76, 77, 80
Cook, Capt. James 47
Cook, Frederick A. 48, 52
Cook Inlet 2, 19, 41, 48, 50, 80, 82, 89, 94
Cooper, William 51, 53
Copper River 11, 19, 49, 50, 71, 73, 75, 76, 100
Cordova 6, 19, 56, 75, 101

Crevasse 16, 17, 19-21, 24, 25, 29-31, 48, 55, 57, 102, 105, 107
Curtis, Edward S. 50

Dall, William H. 49
Darack, Ed 24-25
Davidge, Ric 2, 11
Denali National Park 2, 48, 87, 100, 101
Dickey, W.A. 49
Disenchantment Bay 49, 69, 70
Distributary 62, 103
Doran, Capt. Peter 77
Downwasting 75, 87, 103
Drift 16, 103

Emerson, Benjamin K. 49
Engeln, O.D. von 50
Erosion 6, 8, 19-21, 23, 45, 107, 108
Erratic 35, 48, 103
Esker 20, 103
Eustacy 45, 103

Fairweather Range 5, 52, 68
Faisal, Prince Mohammed al 11
Field, William Osgood Jr. 52, 53
Fiord 20-22, 45, 59-62, 67, 69, 71, 76-78, 94, 101, 103, 108
Firn 14, 15, 56, 64, 103
Firn line 38, 39, 104
Fleming, Mike 2, 9
Foliation 14, 84, 104
Fossils 34-37
Franklin Mountains 8, 97
Frederick Sound 60, 101

"Galloping Glacier" 87
Gannett, Henry 49
Garcia, Rachel 2
Gareloi Island 92, 93
Gerdine, T.G. 84
Gerhard, Bob 91
Gilbert, Grove Karl 49

Glacial stream 14, 16, 20, 28, 103, 104
Glacier Bay 5, 8, 21, 22, 30, 46-50, 52, 66-68, 100, 101, 103
Glacier ice harvesting 2, 7, 10
Glacier travel 24, 25, 57
Glaciers
Agassiz 70
Aialik 81
Allen 75, 76
Amherst 77
Baird 46, 60, 61
Baltimore 77
Barnard (College Fiord) 71, 77
Barnard (St. Elias Mountains) 2, 71, 106
Barry 51, 76, 77
Bartlett 80
Beloit 80
Bering 5, 8, 27, 29, 30, 46, 49, 52-56, 73, 75, 76, 100-102, 106
Black Rapids 30, 84, 87
Blackstone 80
Blockade 90
Brady 69
Brooks 89
Brown 61
Bryn Mawr 76, 77
Burns 80
Byron 56
Canwell 84
Capps 87, 89
Carl 82
Cascade 69, 77
Casement 103, 107
Castner 84
Cathedral 55
Chamberlin 98
Charley 67
Chenega 55, 81, 104
Cheshnina 73
Chickaloon 82
Chickamin 59
Chikuminuk 94
Childs 75, 76, 100, 101
Chilkat 64
Chisana 73
Clark 67
Colony 78
Columbia 51, 53, 76-78
Cone 92
Copper 73
Coxe 77
Crab 92

Crescent 77
Dartmouth 77
Dawes 61
Denver 64
Dixon 81
Doroshin 81
Eagle (Juneau Icefield) 52, 64
Eldridge 87
Eliot 77
Ellsworth 81
Excelsior 2, 81, 108
Exit 5, 40, 100, 101
Fairweather 52, 54
Ferebee 64
Finger 52, 92
Fog 92
Fourpeaked 91
Gilman 67
Goodwin 75
Grand Pacific 50, 67, 69
Grand Plateau 5, 69, 107
Grand Union 95
Grewingk 81
Grinnell 75, 76
Gulkana 53, 84, 87
Guyot 70, 71
Haenke 70
Hallo 91
Harpoon 92
Harriman 76, 77, 106
Harvard 29, 76, 77
Hayes 89
Heney 75, 76
Herbert 52, 64, 105
Hidden 70
Hole-in-the-Wall (Chugach Mountains) 73
Hole-in-the-Wall (Juneau Icefield) 62
Holgate 81
Holyoke 77
Hoonah 67
Hubbard 18, 27, 49, 69, 70
Irene 64
Island 92
Jarvis 84
John 67
Johns Hopkins 67
Johnson (Aleutian Range) 90
Johnson (Alaska Range) 84
Kahiltna 25, 87, 101
Kashoto 67

Kennicott 51, 73
Klutina 78
Knik 78, 80
Koniag 94
Kuskulana 73
Lafayette 77
Lake George 78
Lamplugh 67, 100, 104
La Perouse 50, 69, 107
Lateral 90
Laughton 26
Lawrence 80
LeConte 60
Lemon Creek 64
Lituya 69
Long 73
Lowell 44, 105
Maclaren 87
Malaspina 5, 6, 8, 27, 30, 32, 45, 49, 50, 52-54, 70, 71, 73, 101, 107, 108
Margerie 22, 67
Marquette 83
Martin River 75, 76
Matanuska 5, 21, 25, 50, 52, 78, 100
McBride 68
McCall 97
McCarty 81
Mead 64
Meares 20, 38, 51, 76, 77
Mendenhall 5, 22, 57, 62, 64, 100, 101
Miles 56, 75, 76
Miller 70
Muir 8, 48, 49, 68
Muldrow 23, 30, 87, 89, 100, 103
Nabesna 71, 73
Nelchina 13
Nizina 73
Norris 28, 57, 62, 64
North Crillon 69
North Dawes 61
Northland 80
Northwestern 81
Nunatak 70
Okpilak 97
Outlet 92
Patterson 60
Petrof 81
Phalarope 97
Portage 5, 11, 56, 80, 81, 100
Portlock 81
Princeton 81

Radcliffe 77
Red 41, 90
Revelation 89
Riggs 68
Ripon 80
Russell 71
Ruth 24, 48, 87
Sanford 73
Sawyer 61, 62
Schwan 76
Serpentine 76
Seward 53, 70
Shakes 7
Shamrock 90
Sheridan 76, 100
Sherman 6, 53, 75, 76
Shoup 5, 51, 104
Skilak 81
Skookum 80
Slide 76
Slim 92
Smith 76, 77
Soule 59
South Sawyer 61, 62, 108
Speel 61
Spencer 80
Spotted 90
Steller 73, 76
Surprise 56, 76, 77
Susitna 84, 87
Taku 62, 64
Talkeetna 82
Taylor 80
Tebenkof 80
Thrush 97
Tokositna 87
Toyatte 67
Trail 80
Traleika 89
Trident 84
Trimble 89
Triumvirate 31, 89, 103
Turner 70
Tustumena 81
Tuxedni 89, 90
Tyndall 71
Umbrella 90
Valdez 51, 78, 100
Variegated 29, 30, 70
Vassar 76, 77
Vaughn Lewis 16
Walsh 71
Wellesley 76, 77
Whittier 80
Wolverine 53, 80
Worthington 5, 59, 78, 100, 101

110 ALASKA GEOGRAPHIC®

Wosnesenski 81
Wright 61
Yahtse 71
Yale 76, 77
Yalik 81
Yanert 87
Yentna-Lacuna 87
Glaciers, anatomy of 13-23
Glaciers, distribution of 8, 9
Glaciers, exploration of 47-50, 52, 77
Glaciers, life on 54-56
Glaciers, numbers of 5, 6, 8
Glaciers, size of 5, 9, 26
Glaciers, types of 26-28
  Alpine 8
  Cirque 2, 5, 6, 8, 26, 27, 48, 82, 94, 97, 98, 102
  Hanging 6, 26, 28, 67, 93, 104, 107
  Piedmont 5, 19, 27, 48, 107
  Polar 26, 28, 31
  Reconstituted 28, 107
  Rock 15, 28, 108
  Temperate 26, 28, 31, 32, 45, 106
  Tidewater 5, 10, 22, 26, 28, 51, 54, 55, 60, 61, 67, 69, 70, 76, 77, 81, 101, 103, 108
  Valley 5, 8, 19, 21, 27, 29, 48, 55, 59, 62, 69-71, 73, 82, 89, 90, 92, 93, 98, 101, 105-108
Grant, U.S. 51, 77
Gulf of Alaska 19, 21, 27, 31, 45, 46, 64, 69, 70, 73, 75, 94

Harding Icefield 27, 80, 81, 101, 105
Hare, Greg 42-44
Harriman Alaska Expedition 49, 52, 77
Harriman, Edward Henry 49, 77
Harriman Fiord 50, 76, 77
Harvard Arm 29, 76
Hayes, C.W. 49
Heinselman, Brian 20
Henderson, Keith 93
Heusser, Calvin J. 53
Hickel, Walter J. 11
Higgins, D.F. 51, 77

Holocene 32, 104, 107
Homer 50, 81
Horn 13, 23, 104
Hubbard, Gardiner G. 18
Hubley, Richard C. 53

Iceberg 7, 10, 11, 13, 15-17, 20-22, 27, 31, 54, 55, 62, 75, 77, 78, 81, 101-104
Ice cap 7, 27, 31, 48, 92, 93, 105
Icefall 16, 30, 31, 57, 104, 105, 107
Icefield 7, 27, 31, 42, 48, 59, 62, 64, 71, 80-82, 90, 94, 105
Ice sheet 26, 27, 31, 45, 46, 48
Ice sizzle (see bergy seltzer)
Ice worm 2, 56
Icy Bay 46, 51, 70, 71, 76, 101
Icy Strait 59, 67, 69
Inside Passage 21, 101
Institute of Polar Studies 53
Isostacy 45, 46, 71, 105

Jokulhlaup 13, 17, 18, 70, 80, 105
Juneau 2, 6, 21, 50, 52, 57, 62, 64, 101
Juneau Icefield 2, 13, 27, 28, 53, 55, 59, 61-64, 100, 101, 105
Juneau Icefield Research Program (JIRP) 64

Kachemak Bay 81
Kadin, Mikhail 48
Kame 19, 20, 105
Katmai National Park 90, 91
Kenai Fjord 5, 10
Kenai Mountains 8, 9, 30, 76, 80, 81, 100
Kenai Peninsula 10, 51, 80
Kettle 19, 105
Kigluaik Mountains 8, 9, 95
Kodiak Island 8, 9, 50, 94
Krimmel, Robert 53
Kuzyk, Gerry 42

Lake Clark National Park 32, 91

Landsat 41, 92, 93
La Perouse, Jean Francois de Galaup de 47, 69
Leffingwell, Ernest 97
Little Ice Age 29, 32, 84, 89, 106
Lituya Bay 47, 48, 52, 69, 70
Loess 34, 106

Malaspina, Capt. Don Alessandro 47
Martin, Lawrence 50, 75-77, 80
Mass balance 38, 39, 84, 97, 106
Matheus, Paul 34-37
Mayo, Larry 53
McAllister, Jim 57
McGimsey, Game 2
Mendenhall, W.C. 49, 84
Miller, Maynard 53, 64
Million Dollar Bridge 75
Moffit, F.H. 84
Mohrwinkel, Bill 25
Moraine 6, 30, 40, 62, 70, 73, 76, 90
  Ablation 23, 75, 78, 82, 89, 106
  Ground 16, 106
  Lateral 8, 19, 20, 31, 78, 106, 108
  Medial 2, 16, 19, 24, 31, 47, 55, 56, 71, 106
  Push 62, 106
  Recessional 17, 64, 106
  Terminal (end) 16, 64, 71, 75, 78, 87, 89, 107
Morse, Fremont 49
Moulin 16, 107
Mount Blackburn 71
Mount Fairweather 22, 27, 64, 68, 101
Mount Foraker 87
Mount Gerdine 31, 88, 89
Mount Hayes 83, 84
Mount Iliamna 41, 88-91
Mount Katmai 90, 91
Mount Logan 6, 64
Mount McKinley (Denali) 23, 25, 27, 36, 49, 52, 82, 84, 85, 87, 101
Mount Redoubt 25, 90, 91
Mount Sanford 71
Mount Spurr 87-89
Mount St. Elias 6, 27, 49, 64

Mount Veniaminof 90, 92, 95
Muir, John 48, 49, 60-62, 64, 67, 76
Muldrow, Robert 49

Nassau Fiord 55, 76
National Geographic Society (NGS) 18, 49, 50
National Science Foundation 53
Neal, Tina 2
Nellie Juan Lagoon 17
Névé 15, 87, 107
Nunatak 5, 27, 31, 56, 105, 107

Ogive 16, 30, 107
Osgood, W.H. 49
Outwash plain 19, 60, 62, 69, 87, 105, 107

Paige, Sidney 51
Palache, Charles 49
Pavlof Volcano 90, 92
Permafrost 31, 37
Petersburg 60, 101
Piedmont lobe 6, 27, 56, 70, 73
Plant succession 39, 41, 46, 52
Pleistocene 8, 20, 27, 31, 32, 34-37, 45, 53, 59, 69, 92, 94, 104, 107
Plucking 19, 20, 23, 107
Port Bainbridge 51, 76
Port Nellie Juan 51, 76
Portland Canal 59, 60
Post, Austin 29, 30, 53
Prince William Sound 5, 17, 21, 47-50, 52, 55, 76, 78, 80, 81, 100, 101
Prokosch, Gary 2, 10
Puget, Lt. Peter 47
Putnam, Robert 93

Rand, Dean 17
Red algae 19, 55
Regelation 28, 107
Reid, Henry Fielding 48, 49, 52
Rennick, Penny 56
Retreat 5, 8, 17, 19, 20, 32, 38, 46, 53, 59-62, 64, 67, 68, 70, 71, 73, 75, 77, 78, 80-82, 84, 87, 89, 91, 92, 94, 97, 98, 102, 105, 106

Revelation Mountains 82, 89
Roach, Angela 2
Roche moutonnée 23, 107
Rock flour 18, 19, 108
Romanzof Mountains 8, 97, 98
Russell Fiord 30, 49, 69, 70
Russell, Israel Cook 18, 49, 70

Sargent Icefield 2, 27, 52, 80, 81, 101, 105
Sargent, R.H. 52
Schafer, Al 2, 10
Schwatka, Frederick 49
Sea level 8, 20, 45, 62, 68-70, 80, 90, 92, 103
Sérac 57, 108
Service, Robert 56
Seward 2, 10, 40, 81, 101
Seward Peninsula 8, 9, 95
Shain, Dan 2, 56
Sharp, Robert 53
Sherman, William Tecumseh 75
Silt 5, 16, 34, 45, 97, 108
Sitka 48, 50, 59
Skagway 26, 59, 64
Skidmore, E.R. 49
Smith, Philip 97, 98
Smithsonian Institution 18, 49
St. Elias Mountains 2, 8, 9, 20, 30, 44, 49, 52, 64, 65, 71, 100, 101
Stephens Passage 60, 61
Stikine Icefield 27, 59-61, 101, 105
Stikine River 11, 48, 49, 60, 61
Suncup 53, 108
Surge 6, 29, 30, 41, 46, 55, 69, 70, 73, 75, 76, 84, 87, 89, 108
Syverson, Rachel and Tasha 39

Talkeetna 82, 101
Talkeetna Mountains 8, 9, 82
Tarn 68, 108
Tarr Inlet 22, 67
Tarr, Ralph 50, 76
Teben'kov, Admiral

Mikhail Dmitrievich 48
Terminus 2, 5, 16, 20, 22, 29, 31, 32, 38, 39, 41, 48, 49, 53, 55, 56, 60-62, 64, 67-71, 73, 75-78, 81, 82, 84, 89, 90, 92, 102, 106-108
Till 16, 18, 26, 106, 108
Tongass National Forest 46, 52
Tordrillo Mountains 31, 82, 87, 89
Trabant, Dennis 2
Tracy Arm 5, 21, 61, 62
Trentier, Kozima 48
Trimline 23, 31, 32, 108

Unakwik Inlet 38, 76, 77
Unalaska Island 92, 93
University of Alaska 39, 53, 73, 84
University of Alaska Museum 37, 44
U-shaped valley 20, 21, 47, 59, 84, 87, 103, 108
U.S. Army Corps of Engineers 14, 93
U.S. Department of Defense 39
U.S. Forest Service 64, 100
U.S. Geological Survey (USGS) 2, 9, 18, 29, 34, 40, 49-51, 53, 78, 84, 93
U.S. Navy 51, 64

Valdez 20, 38, 59, 78
Vancouver, Capt. George 47, 67, 69, 70
Verhey, John 7

Washburn, Bradford 52, 71
Westdahl, Ferdinand 49
Whidbey, James 67
White, Stroller 56
Wisconsinan Glaciation 31, 32, 35, 36
Wood River Mountains 8, 9, 94
World Data Center for Glaciology 53
Wrangell 7, 50, 60
Wrangell Mountains 8, 9, 30, 49, 71-75

Yakutat 19, 30, 49, 50, 101

Yakutat Bay 5, 48-50, 52, 65, 69, 101
Young, Samuel Hall 48
Yukon Territory 42, 43
Yunaska Island 92, 93

Zone of brittle flow (fracture) 24, 29, 30
Zone of plastic flow 29

PHOTOGRAPHERS

Baker, Bruce 57
Beeman, Marydith 35
Bol, Tom 25, 53
Bowers, Harvey 56
Butterfield, Bob 15, 26, 58, 84, 102, 108
Cornelius, Don 60
Corral, Roy M. 11, 108
Davis, James L. 43
Endres, Patrick J. 17, 20
Folkens, Pieter 46, 62, 108
Frederick A. Cook Society 36
Government of Yukon Territory 42
Hirschmann, Fred 1, 12, 21, 31, 50, 51, 68, 86, 87, 89, 91, 103, 104
Hyde, John 13, 21, 22, 28, 52, 104
Jans, Nick 5, 104
Johnson, R.E. 4, 6, 7, 18, 30, 49, 59, 107, 108
Lotscher, Chlaus 33
McCullough, Brian 82
Metzler, Curvin Cover, 48, 73, 108
Molnia, Bruce 8, 14, 16, 19 (2), 27, 29, 30, 31, 38, 41, 46, 53, 55, 64, 69, 70, 71, 75, 81, 102, 103, 104, 105, 106, 107, 108
Neal, Tina (AVO) 95
Nickles, Jon R. 40, 47, 77
Okonek, Brian 37, 52, 98
Rennick, Penny 10
Rose, Hugh S. 23, 54, 78, 97, 99, 103, 104, 105
Schwieder, John 101
Speaks, Michael R. 44, 105
Syverson, Greg 39
Taft, Loren 45, 76
Walker, Harry M. 22, 24, 67

ALASKA GEOGRAPHIC® Back Issues

The **North Slope**, Vol. 1, No. 1. Out of print.
One Man's Wilderness, Vol. 1, No. 2. Out of print.
Admiralty...Island in Contention, Vol. 1, No. 3. $9.95.
Fisheries of the North Pacific, Vol. 1, No. 4. Out of print.
Alaska-Yukon Wild Flowers, Vol. 2, No. 1. Out of print.
Richard Harrington's Yukon, Vol. 2, No. 2. Out of print.
Prince William Sound, Vol. 2, No. 3. Out of print.
Yakutat: The Turbulent Crescent, Vol. 2, No. 4. Out of print.
Glacier Bay: Old Ice, New Land, Vol. 3, No. 1. Out of print.
The Land: Eye of the Storm, Vol. 3, No. 2. Out of print.
Richard Harrington's Antarctic, Vol. 3, No. 3. $9.95.
The Silver Years, Vol. 3, No. 4. $19.95. Limited.
Alaska's Volcanoes, Vol. 4, No. 1. Out of print.
The Brooks Range, Vol. 4, No. 2. Out of print.
Kodiak: Island of Change, Vol. 4, No. 3. Out of print.
Wilderness Proposals, Vol. 4, No. 4. Out of print.
Cook Inlet Country, Vol. 5, No. 1. Out of print.
Southeast: Alaska's Panhandle, Vol. 5, No. 2. Out of print.
Bristol Bay Basin, Vol. 5, No. 3. Out of print.
Alaska Whales and Whaling, Vol. 5, No. 4. $19.95.
Yukon-Kuskokwim Delta, Vol. 6, No. 1. Out of print.
Aurora Borealis, Vol. 6, No. 2. $19.95.
Alaska's Native People, Vol. 6, No. 3. $29.95. Limited.
The Stikine River, Vol. 6, No. 4. $9.95.
Alaska's Great Interior, Vol. 7, No. 1. $19.95.
Photographic Geography of Alaska, Vol. 7, No. 2. Out of print.
The Aleutians, Vol. 7, No. 3. Out of print.
Klondike Lost, Vol. 7, No. 4. Out of print.
Wrangell-Saint Elias, Vol. 8, No. 1. Out of print.
Alaska Mammals, Vol. 8, No. 2. Out of print.
The Kotzebue Basin, Vol. 8, No. 3. Out of print.
Alaska National Interest Lands, Vol. 8, No. 4. $19.95.
***Alaska's Glaciers**, Vol. 9, No. 1. Rev. 1993. $21.95. Limited.
Sitka and Its Ocean/Island World, Vol. 9, No. 2. Out of print.
Islands of the Seals: The Pribilofs, Vol. 9, No. 3. $9.95.

Alaska's Oil/Gas & Minerals Industry, Vol. 9, No. 4. $9.95.
Adventure Roads North, Vol. 10, No. 1. $9.95.
Anchorage and the Cook Inlet Basin, Vol. 10, No. 2. $19.95.
Alaska's Salmon Fisheries, Vol. 10, No. 3. $9.95.
Up the Koyukuk, Vol. 10, No. 4. $9.95.
Nome: City of the Golden Beaches, Vol. 11, No. 1. $19.95.
Alaska's Farms and Gardens, Vol. 11, No. 2. $19.95.
Chilkat River Valley, Vol. 11, No. 3. $9.95.
Alaska Steam, Vol. 11, No. 4. $19.95.
Northwest Territories, Vol. 12, No. 1. $9.95.
Alaska's Forest Resources, Vol. 12, No. 2. $9.95.
Alaska Native Arts and Crafts, Vol. 12, No. 3. $24.95.
Our Arctic Year, Vol. 12, No. 4. $19.95.
***Where Mountains Meet the Sea**, Vol. 13, No. 1. $19.95.
Backcountry Alaska, Vol. 13, No. 2. $9.95.
British Columbia's Coast, Vol. 13, No. 3. $9.95.
Lake Clark/Lake Iliamna, Vol. 13, No. 4. Out of print.
Dogs of the North, Vol. 14, No. 1. Out of print.
South/Southeast Alaska, Vol. 14, No. 2. $21.95. Limited.
Alaska's Seward Peninsula, Vol. 14, No. 3. $19.95.
The Upper Yukon Basin, Vol. 14, No. 4. $19.95.
Glacier Bay: Icy Wilderness, Vol. 15, No. 1. Out of print.
Dawson City, Vol. 15, No. 2. $19.95.
Denali, Vol. 15, No. 3. $9.95.
The Kuskokwim River, Vol. 15, No. 4. $19.95.
Katmai Country, Vol. 16, No. 1. $19.95.
North Slope Now, Vol. 16, No. 2. $9.95.
The Tanana Basin, Vol. 16, No. 3. $9.95.
***The Copper Trail**, Vol. 16, No. 4. $19.95.
***The Nushagak Basin**, Vol. 17, No. 1. $19.95.
***Juneau**, Vol. 17, No. 2. Out of print.
***The Middle Yukon River**, Vol. 17, No. 3. $19.95.
***The Lower Yukon River**, Vol. 17, No. 4. $19.95.
***Alaska's Weather**, Vol. 18, No. 1. $9.95.
***Alaska's Volcanoes**, Vol. 18, No. 2. $19.95.
***Admiralty Island: Fortress of Bears**, Vol. 18, No. 3. Out of print.
Unalaska/Dutch Harbor, Vol. 18, No. 4. Out of print.
***Skagway: A Legacy of Gold**, Vol. 19, No. 1. $9.95.
Alaska: The Great Land, Vol. 19, No. 2. $9.95.
Kodiak, Vol. 19, No. 3. Out of print.
Alaska's Railroads, Vol. 19, No. 4. $19.95.
Prince William Sound, Vol. 20, No. 1. $9.95.
Southeast Alaska, Vol. 20, No. 2. $19.95.
Arctic National Wildlife Refuge, Vol. 20, No. 3. $19.95.
Alaska's Bears, Vol. 20, No. 4. $19.95.
The Alaska Peninsula, Vol. 21, No. 1. $19.95.
The Kenai Peninsula, Vol. 21, No. 2. $19.95.
People of Alaska, Vol. 21, No. 3. $19.95.
Prehistoric Alaska, Vol. 21, No. 4. $19.95.
Fairbanks, Vol. 22, No. 1. $19.95.
The Aleutian Islands, Vol. 22, No. 2. $19.95.
Rich Earth: Alaska's Mineral Industry, Vol. 22, No. 3. $19.95.
World War II in Alaska, Vol. 22, No. 4. $19.95.
Anchorage, Vol. 23, No. 1. $21.95.
Native Cultures in Alaska, Vol. 23, No. 2. $19.95.
The Brooks Range, Vol. 23, No. 3. $19.95.
Moose, Caribou and Muskox, Vol. 23, No. 4. $19.95.

Alaska's Southern Panhandle, Vol. 24, No. 1. $19.95.
The Golden Gamble, Vol. 24, No. 2. $19.95.
Commercial Fishing in Alaska, Vol. 24, No. 3. $19.95.
Alaska's Magnificent Eagles, Vol. 24, No. 4. $19.95.
Steve McCutcheon's Alaska, Vol. 25, No. 1. $21.95.
Yukon Territory, Vol. 25, No. 2. $21.95.
Climbing Alaska, Vol. 25, No. 3. $21.95.
Frontier Flight, Vol. 25, No. 4. $21.95.
Restoring Alaska: Legacy of an Oil Spill, Vol. 26, No. 1. $21.95.
World Heritage Wilderness, Vol. 26, No. 2. $21.95.
The Bering Sea, Vol. 26, No. 3. $21.95.
Russian America, Vol. 26, No. 4, $21.95
Best of *ALASKA GEOGRAPHIC*®, Vol. 27, No. 1, $24.95
Seals, Sea Lions and Sea Otters, Vol. 27, No. 2, $21.95
Painting Alaska, Vol. 27, No. 3, $21.95
Living Off the Land, Vol. 27, No. 4, $21.95
Exploring Alaska's Birds, Vol. 28, No. 1, $23.95

* Available in hardback (library binding) — $24.95 each.

PRICES AND AVAILABILITY SUBJECT TO CHANGE

Membership in The Alaska Geographic Society includes a subscription to *ALASKA GEOGRAPHIC*®, the Society's colorful, award-winning quarterly. Contact us for current membership rates or to request a free catalog.

The *ALASKA GEOGRAPHIC*® back issues listed above can be ordered directly from us. **NOTE:** This list was current in mid-2001. If more than a year has elapsed since that time, contact us before ordering to check prices and availability of back issues, particularly for books marked "Limited."

When ordering back issues please add $5 for the first book and $2 for each additional book ordered for Priority Mail. Inquire for postage rates to non-U.S. addresses. To order, send check or money order (U.S. funds) or VISA or MasterCard information (including expiration date and your daytime phone number) with list of titles desired to:

ALASKA GEOGRAPHIC

P.O. Box 93370 • Anchorage, AK 99509-3370
Phone (907) 562-0164 • Toll free (888) 255-6697
Fax (907) 562-0479 • e-mail: info@akgeo.com

NEXT ISSUE: Vol. 28, No. 3

Inupiat and Yupik People of Alaska

The Inupiat and Yupik people inhabit northern and western Alaska, where they blend their ancestral traditions with economic and cultural realities of the modern world. Well-known Native writers Susie Silook and Jana Harcharek, and anthropologist Ann Fienup-Riordan examine the daily lifestyle, the culture, and the social history of these people. To members fall 2001.

Alaska in Maps: A Thematic Atlas, CD-ROM

This exciting, visually rich CD-ROM teaches geography and Geographic Information Systems easily. Use the interactive map section to make your own thematic maps. Place communities on top of Native corporation lands. Overlay mineral deposits on a transportation map. The resulting maps can be used with PowerPoint and other software. For PC; call for system requirements. To order, send $35 plus $3 shipping/handling for each CD to:

$35 (+$3 S/H)

ALASKA GEOGRAPHIC
P.O. Box 93370-CDG • Anchorage AK 99509
(907) 562-0164 • Toll free (888) 255-6697